MANUFACTURE

OF

METALLIC ALLOYS.

A PRACTICAL GUIDE

FOR THE

MANUFACTURE

OF

METALLIC ALLOYS:

COMPRISING THEIR

CHEMICAL AND PHYSICAL PROPERTIES,

WITH THEIR

PREPARATION, COMPOSITION, AND USES.

TRANSLATED FROM THE FRENCH OF

A. GUETTIER,

ENGINEER AND DIRECTOR OF FOUNDRIES, AUTHOR OF "LA FONDERIE EN FRANCE," ETC. ETC.

BY

A. A. FESQUET

CHEMIST AND ENGINEER.

PHILADELPHIA:
HENRY CAREY BAIRD
INDUSTRIAL PUBLISHER,
406 Walnut Street.
LONDON:
SAMPSON LOW, SON, & MARSTON,
CROWN BUILDINGS, 188 FLEET ST.
1872.

PHILADELPHIA:
COLLINS, PRINTER, 705 JAYNE STREET.

INTRODUCTION.

HUNDREDS of times have we held back from the undertaking of a special treatise on alloys.

A complete work, adequate to the importance of the subject, would require innumerable researches and studies, and one volume would not be sufficient. Yet, it would scarcely be possible to give anything beyond a concise idea of a subject entirely too vast and complex to be treated in a strict and exact manner.

Let us consider all the metals actually employed, from those which are essentially industrial, to the precious metals which belong to the arts rather than to industry proper; and up to those modern metals so little known that they still remain exclusively within the limits of scientific investigation. When we see these various metals combining with each other, one by one, two by two, three by three, &c., and in various proportions, we may well ask if it be possible to create a methodical and absolute treatise on alloys.

Not only would it be impossible to resolve all of the problems arising out of the multiple combinations of the metals with each other, on account of their innumerable quantity; but, as experience must be the definitive test, it is impossible for most of these problems to be solved without practical studies, which alone are capable of throwing sufficient light upon this subject.

In order to study all of the alloys which may be produced by the various metals, beginning with the usual ones and finishing with the new ones, a considerable expenditure of time and money would be necessary. A lifetime would scarcely be sufficient for producing and studying with profit all of the elementary combinations which the question requires.

Few, if any, persons among those interested in the metallurgic art, have made longer and more complete researches on alloys than we have. However, with entire humility, we are ready to acknowledge that our efforts, which may have aided the industry up to a certain point, are far from having elucidated the least complex parts of the question.

We have endeavored to give to these studies a practical turn, by considering the alloys according to their conspicuous qualities; and following the successive variations in combination of the common metals, and the part borne by each one in these modifications. But these researches, already protracted and difficult, have touched only one part of our intended programme.

We have been obliged to give approximate results, in place of precise numbers, for the part played by each metal in regard to the resistance, hardness, specific gravity, fusibility, &c. of alloys. But, to have done otherwise, it would have been necessary to multiply the experiments and the verifications, and to have mechanical trials intervening in a question where the principal part is the work of the founder.

Time and opportunities have failed, not only for completing these first studies, but also for beginning new ones. Nor can we say when we shall resume this question, if at all.

Then, and until others more successful or better endowed increase the knowledge of alloys by new and correct data, there is nothing left but to sum up, as clearly and as briefly as possible, all that has been ascertained in regard to alloys, by others and by ourself.

On that account, and in order to make a book within the means of all workers, we shall only examine the combinations of the most usual metals.

The known metals may be subdivided into four distinct classes:—

1st. The metals especially industrial, that is to say, those which are most in use in all kinds of manufactures. They are: Copper, tin, zinc, lead, iron, steel, &c.

2d. The metals which belong to the arts, but whose importance is secondary. These are: Bismuth, antimony, nickel, arsenic, and mercury.

3d. The precious metals which belong to the arts, or more particularly to the manufacture of objects of luxury. These are: Gold, silver, aluminium, and platinum.

4th. The metals scarcely used in industry or in

alloys; most of them being, at present, without any clearly demonstrated usefulness.

After some preliminary explanations about the physical and chemical properties of the metals and alloys, we shall examine the metals of the first class in view of their mutual combinations. This investigation is a sort of commentary upon the results of our personal researches which were published a few years ago, under the title of *Recherches sur les Alliages des Métaux industriels.*

This portion will be followed by general indications concerning the metals of the second and third classes, in view of the alloys with themselves and with metals of the other classes; most of these metals, with a few well-known exceptions, having given rise to observations more curious and scientific than practical and useful.

Lastly, we shall consider the metals of the fourth class only in regard to their possible association with alloys presenting certain interest in the arts.

If we add to these data concise observations in relation to the composition and preparation of the mixtures, to their smelting and moulding, &c.—in one word, to the industrial treatment of alloys—and if we annex to that the series of compositions of alloys which have been found practical and useful in various sorts of manufactures, we shall have composed a treatise on alloys, or an experimental guide, which will present in a concise form the principal elements of this important question; but we shall still be far from having elucidated even a small portion of a subject which, in many respects, demands the revelations of science combined with a large experience.

For instance, when the new metals, comparatively unknown, shall be added to the usual metals whose alloys have been tested by long practice, who can foresee the results of these new combinations, or the new qualities imparted to the ancient metals, as has been done, with more or less success, to copper by aluminium, and to iron and steel by wolfram (tungsten)?

In regard to the ordinary metals, whose principal combinations are well known, we have to ascertain the proportions, the elements best adapted to certain uses, the hardness and malleability, &c.; and to educe scientifically with figures these proportions and elements, and to cause them to rise above the empiric state in which they have lingered under the rules of practical routine. This, above all, is the aim toward which our efforts must tend.

With those new metals which are not well known, we must endeavor, by uniting them with known alloys, to produce new combinations, which may prove real revelations, and by which the science of alloys will have made, in a short time, very rapid and unexpected strides. This is the road to sure progress, and for improvements in the working and employment of metals.

Because it is possible to unite in indefinite proportions some metals, which, being thoroughly mixed during their fusion, remain so after solidification, we must not infer that all alloys are mixtures only. Metals, equally with all other chemical substances, combine in definite proportions, the limits of which must be known, if we desire to obtain an intimate and normal union. Indeed, our object is not to create alloys with any proportions or metals which, by liquation, will not produce homogeneous castings. If such were the case, the different parts of the castings would have different compositions, in indefinite proportions.

Therefore the science of alloys is not a mere guess-work, which consists in taking metals, no matter what they be, and in mixing them without rule or measure. We must use those quantities best adapted to such and such metals, which we intend to use in an alloy; and it sometimes happens that a very small proportion of a given metal will impart to another metal new and unexpected properties.

This is a reason why the study of alloys made with certain metals, which at the present time have been but little experimented upon, may produce very important results; and we cannot too strongly recommend such researches to those of our readers who may attempt industrial experiments in the department of metallic alloys.

A. G.

CONTENTS.

PART I.

CHAPTER I.

GENERAL OBSERVATIONS ON THE METALS WHICH ARE COMMONLY USED FOR ALLOYS.

CHAPTER II.

PHYSICAL AND CHEMICAL PROPERTIES OF ALLOYS.

CHAPTER III.

PREPARATION AND COMPOSITION OF ALLOYS.

PART II.

CHAPTER I.

ALLOYS OF THE METALS MOST USED IN THE ARTS.

CHAPTER II.

ALLOYS OF THE METALS OF SECONDARY IMPORTANCE IN THE ARTS.

CHAPTER III.

ALLOYS OF THE PRECIOUS METALS, BELONGING ESPECIALLY TO THE ARTS OF LUXURY.

CHAPTER IV.

ALLOYS OF THE METALS RARELY OR NEVER USED IN THE ARTS.

PART III.

CHAPTER I.

BRONZES OF ART.

CHAPTER II.

ALLOYS FOR COINAGE.

CHAPTER III.

ALLOYS FOR PIECES OF ORDNANCE, ARMS, PROJECTILES, ETC.

CHAPTER IV.

ALLOYS FOR ROLLING AND WIRE DRAWING.

CHAPTER V.

COPPER ALLOYS FOR SHIP SHEATHINGS.

CHAPTER VI.

ALLOYS FOR TYPE, ENGRAVING PLATES, ETC.

CHAPTER VII.

ALLOYS FOR BELLS, MUSICAL INSTRUMENTS, ETC.

CHAPTER VIII.

ALLOYS FOR PHILOSOPHICAL AND OPTICAL INSTRUMENTS.

CHAPTER IX.

ALLOYS FOR JEWELRY, GOLD AND SILVER WARES, BRITANNIA WARE, ETC.

CHAPTER X.

WHITE ALLOYS.

CHAPTER XI.

FUSIBLE ALLOYS.

CHAPTER XII.

ALLOYS FOR MACHINERY, ANTI-FRICTION METALS, ETC.

CHAPTER XIII.

SOLDERS.

CHAPTER XIV.

MISCELLANEOUS ALLOYS.

PRACTICAL GUIDE

FOR THE

MANUFACTURE OF METALLIC ALLOYS.

PART I.

I.

GENERAL OBSERVATIONS ON THE METALS WHICH ARE COMMONLY USED FOR ALLOYS.

THE metals which we are about to consider are those of the first three classes, as indicated in our introduction.

These metals, whatever be their value or usefulness, are entitled to a certain degree of importance in manufactures.

Although some of them have been long known and some are modern, all have been sufficiently well studied; and it is not necessary for us to point out all their acknowledged characteristics.

At every epoch in the study of metals, recourse has been had, as at present, to certain combinations which exhibit their usefulness in every respect.

Used in the pure state, that is to say, without having been alloyed with other metals which would impart to them particular qualities, these metals would have few applications in industry; we must, however, except iron, which by itself may be applied to innumerable uses.

On the contrary, when forming some of those thousands of combinations resulting from their union with each other, certain metals, such as copper, tin, and lead, which by themselves would be of secondary interest in the arts, acquire an enormous importance as soon as they are alloyed.

Thence we see all the interest attached to the study of alloys, which requires the aid both of science and practice for improvement and progress.

However, it is necessary that all of our readers who are interested in this study should have presented to them some general data concerning the characteristics and properties of the metals which are the component parts of alloys.

We admit that most of our readers possess this information: but memory might fail some of them, and some essential though elementary details may escape others. Nevertheless, a book like this should be complete, and it ought to include all the rudiments absolutely necessary for the understanding of the subject, without the trouble of searching for the information in other books.

The metals which we are about to consider are:—

Copper.
Tin.
Zinc.
Lead.
Iron.
Bismuth.
Antimony.
Nickel.
Arsenic.
Mercury.
Gold.

Silver.
Platinum.
Aluminium.

We shall give a cursory glance at each of these metals, in the order in which they have been named.

Copper.

Copper is one of the oldest known metals. Its color is brown pink or a brilliant brown red, and presents shades varying from yellow to red, according to the purity of the metal. A good ingot copper has a metallic appearance with a bright and regular glitter, and *without brown or black spots; its grain is fine, close,* without hard portions, and is easily abraded by the file.

The specific gravity of copper varies between 8.8 and 8.9. It is feebly sonorous, and its smell and savor are little appreciable, but very unpleasant. It is malleable, ductile, and tenacious. Strongly heated, although but slightly volatile, it gives off a fine green vapor. Heated in contact with the air, it readily becomes largely oxidized, and loses part of its ductility and malleability. Exposure to a damp atmosphere produces on its surface a greenish pellicle of an oxide called verdigris. It is attacked more or less rapidly by acids, and is easily dissolved by nitric acid.

Copper may be readily alloyed with other metals; except iron and lead, the alloys with which are difficult to form.

Tin.

Tin appears to be the oldest metal employed in the arts, and is mentioned in the history of the earliest ages. White, and with a lustre nearly as brilliant as that of silver, it tarnishes more easily and rapidly than the latter metal.

Its specific gravity varies between 7.3 and 7.5, whether it is cast, hammered, or laminated.

Tin, when bent, produces a peculiar crackling noise, which may be made use of for ascertaining the purity of the metal. Certain sorts of tin are pure, such as the Banca, Straits, or Malacca ingots, as were also some English marks, which are now seldom found in the trade in a pure state. Those tins which are adulterated with foreign metals, such as lead, iron, copper, and arsenic, may be recognized not only by a difference in the crackling noise, but also by a dull appearance and a more or less radiated surface, according to the greater or less quantity of foreign matters. The smell and savor of tin are very perceptible and unpleasant. This metal is tenacious, ductile, and very malleable; when pure, it is very soft, but less so than lead. Without being volatile, it is rapidly oxidized when kept for a long time in a state of fusion, with free access to the air. It is corroded by acids, which, acting upon its surface, produce a metallic crystalline appearance.

It is decomposed by nitric, sulphuric, and muriatic acids, and may be combined and alloyed with most of the metals employed in manufactures.

ZINC.

Zinc, sometimes called *Spelter*, was possibly employed by the ancients in the state of alloy, by combining its ores with copper, tin, and lead; but as a metal it was not known until a long time after the metals we have just named. Even as regards its uses in industry, zinc has been employed only since the beginning of this century.

Zinc is bluish-white, and the color of its surface is

similar to that of lead. It has a crystalline fracture with large radiating laminæ, which tarnish in the air.

Its specific gravity varies between 6.9 and 7.2.

Very malleable at a temperature ranging between 120 and 150 degrees centigrade, it is very brittle beyond these limits. At about 300° C. it becomes so brittle that it is possible to pulverize it. Compared with other metals, zinc is soft and possesses little tenacity; it is not sonorous, and its smell and savor are peculiar, although not very perceptible.

When melted, zinc is quickly oxidized by air; and, if the temperature is raised above that of fusion, it will volatilize rapidly and its vapors will burn, producing a flaring light and white fumes much like cotton flakes.

By the action of the air, zinc is easily oxidized at first; but soon the oxidation ceases.

Acids, even diluted, attack zinc rapidly. Caustic alkalies will also oxidize and dissolve it. This metal may be alloyed with most of the usual metals.

Lead.

Lead, a metal known to the ancients at the same time as copper and tin, is bluish-white, has a very brilliant lustre when freshly cut, but becomes quickly tarnished. Malleable and ductile, this metal possesses little tenacity; without savor, it has a sensible and peculiar odor. It is so soft that it may be scratched by the nail, and leaves a gray streak when rubbed against wood, metals, and paper. Its specific gravity is 11.445. It tarnishes rapidly in contact with the air, and becomes covered with a dark pellicle, which, after a certain lapse of time, turns grayish-white. When melted, it may be rapidly oxidized, if it is stirred, and the air has free access to the surface of the molten metal. The more the temperature is increased, the more rapidly the oxidation goes on. At a red heat, lead burns with a flame of a livid white. Nitric acid and aqua regia, even when diluted, attack it easily. Sulphuric and muriatic acids have little action upon it, when cold.

Lead may be alloyed with most of the metals. However, such alloys are difficult to form when the specific gravities and temperatures of fusion of the other metals are very different from those of its own. Lead has a great affinity for gold and silver. Industry utilizes this property for separating, by cupellation, gold and silver from the other metals and earths which accompany them.

IRON.

Although *Iron* was well known to the ancients, it is only in modern times that its production and use began to be developed. This metal, which at present, in its various states of cast iron, wrought iron, and steel, is foremost among the metals employed in the arts, has received a prodigious development, mostly in the present century.

It is bluish-gray or grayish-white when granular or laminated, and its lustre is bright or dull, according as it has been drawn or cast. Its hardness, tenacity, ductility, and malleability vary also with its various states. Cast or raw iron is hard and brittle, whereas wrought iron and steel are exceedingly resisting, malleable, and ductile. The specific gravity of pig-iron is 7.20, and that of iron or steel rises to 7.7 and 7.9.

Iron is very easily oxidized; in a damp atmosphere the rust has a very destructive action, and necessitates the employment of varnishes and other preservative coatings. In the molten state, or at a red heat, iron, when in contact with the air, is rapidly oxidized. Acids attack and dissolve it easily; and this metal, notwithstanding its qualities point to a great stability and durability, requires to have its outer surface protected against destructive agents.

Iron does not alloy well with most of the metals; a peculiar state and the high temperature necessary for its fusion, etc., are hindrances to its being easily alloyed.

Bismuth.

Bismuth does not seem to have been known by the ancients. Agricola is the first author who mentions it, in a book published in 1546. The discovery of this metal appears, therefore, to date from the sixteenth century.

Bismuth has a grayish-white color, shading to that of red. Its fracture is lamellar, and it possesses neither smell nor savor. Its specific gravity varies between 9.83 and 9.89. This metal, as found in the trade, is brittle, with little tenacity, and without any ductility or malleability.

Of all metals, bismuth possesses the greatest facility for crystallizing. When cooled slowly, the crystals it produces are remarkable by their size, their cubic shape, and their peculiar lustre.

This metal is very fusible, volatile, and oxidizable at a high temperature, like many metals which are not refractory. In a damp atmosphere it becomes covered with a reddish-brown pellicle of oxide. At a red heat it burns with a bluish flame, and produces fumes of a yellow-red color.

The high price of bismuth limits its uses. This metal is mostly employed for fusible alloys and those of typography, where the metals usually combined with it are lead, tin, antimony, etc.

Antimony.

Antimony is of relatively limited use in industry, except for certain special alloys. Its color is silver-white, shading to a bluish-white; its fracture is entirely lamellar, and it is so brittle and friable that it can be easily pulverized. According to its degree of purity, its specific gravity varies from 6.65 to 6.85.

Antimony melts at a temperature below that of a red heat, and fills the air with thick white fumes. Diluted or concentrated nitric acid attacks it, and allows its separation, whether from its ores or from its alloys. However, these alloys are few, and used principally for printing-types, plates for engraving music, and certain compounds of lead, tin, and antimony, to which small quantities of copper and bismuth are sometimes added.

Antimony is employed in medicine and pharmacy. In the treatment of metals, it is used in the metallic state, and is generally known under the name of regulus of antimony. Gold, when exposed to the vapors of antimony, immediately loses its ductility and malleability, and becomes as brittle as antimony itself.

Nickel.

Nickel was discovered by Cronstedt at about the middle of the eighteenth century. It has a grayish-white color nearly like that of platinum, and its fracture is crooked. Its specific gravity varies between 8.4 and 8.8, according to the degree of compression it has been subjected to. Worked when hot, it takes a fibrous structure, and may be forged and laminated. Its hardness is very nearly that of iron; and it may be easily polished, acquiring a great brightness by this operation.

Nickel does not oxidize or tarnish at the ordinary temperature; even when hot, it is slowly and with difficulty that it becomes oxidized. This property furnishes a reason for several countries having introduced the use of nickel in the manufacture of small coin.

Nickel alloys very well with copper, tin, zinc, antimony, iron, cobalt, etc., and is especially employed for those alloys which imitate or replace silver.

Arsenic.

Arsenic, which chemists place among the metalloids, possesses in the metallic state a steel-gray color, which quickly tarnishes. It is seldom employed in this form. It is very brittle, fuses readily, and is then immediately volatilized, unless the fusion be effected in closed vessels. Heated in contact with the air, it burns with a blue flame, emits a garlic odor, and becomes converted into a white volatile substance, which is the white arsenic, or arsenious acid. White arsenic is more soluble in hydrochloric acid than in water. Its uses are for a few pharmaceutical preparations, the manufacture of fine glass, such as Bohemian glassware, that of Sheele's green, and of other greens employed in dyeing.

Arsenic is rarely alloyed. However, it is employed in the composition of telescope mirrors, and of some other metallic combinations which are seldom used, and which will be noticed hereafter. Its specific gravity is 5.63.

Mercury.

Mercury, sometimes called *Quicksilver*, is as bright and nearly as white as silver. Fluid at the ordinary temperature, it becomes solid at 39½° C. below the freezing point. In this state it possesses some tenacity and malleability. Liquid mercury has neither taste nor smell. It transmits heat well, and expands considerably. It does not "wet," that is to say, has no molecular attraction for many substances. Its specific gravity when solid is 14.39, when fluid 13.60, and in vapor 6.976.

Heated in contact with the air, at from 350° to 360° C., which is nearly its point of ebullition, it is transformed into a red oxide.

Like porous substances, mercury absorbs a certain quantity of air and dampness, which cannot be expelled except by ebullition. Everybody knows the sensation of burning produced by the melting of solid mercury in contact with a portion of the human body; also the disorders it occasions when introduced into the human economy. We shall not enlarge on these phenomena, which are foreign to this book.

Most acids are without action upon this metal, although it is dissolved, with evolution of sulphurous and nitrous fumes, by concentrated sulphuric or nitric acid.

In the metallic state, mercury is employed in pharmacy; in the construction of barometers, thermometers, and manometers; in tinning looking glasses; in amalgamating silver and gold; in producing various colors for the arts; in the manufacture of the fulminate for percussion-caps, etc.

Its alloys, which bear the name of *Amalgams,* are formed with nearly all metals, especially with copper, lead, zinc, tin, bismuth, silver, and gold. It does not amalgamate, or rather combines with difficulty, with iron, nickel, platinum, cobalt, manganese, etc.

GOLD.

Gold is one of the metals known in the earliest ages. Its precious qualities of unalterability, ductility, and rarity have made it the most valuable metal from the beginning of the world.

Gold is of a fine yellow, somewhat reddish color.

It has neither smell nor taste; it is the most ductile, malleable, and the least oxidizable of all metals. Its specific gravity varies from 19.26 to 19.37, whether melted or laminated or hammered.

Nitric, hydrochloric, and sulphuric acids do not attack it; but it is dissolved by aqua regia (a mixture of nitric and hydrochloric acids), and by the alkaline polysulphides. At a very high temperature, gold is volatilized with a green flame.

The alloys of gold would be easy with most metals; however, they are limited on account of the price of gold, and, therefore, are only those where gold is the essential portion of the alloy.

SILVER.

Silver, which ranks next to gold among precious metals, has an origin and uses which are not so old as those of gold, although dating from an early age.

Its texture is of a dead white color, which will receive a brilliant polish. On account of its malleability, ductility, and resistance to oxidation, it, like gold, is one of the most precious and remarkable metals.

Its specific gravity varies from 10.47 to 10.45, according to the treatment to which it has been previously submitted.

Unacted upon by air alone, silver, under the influence of a very great heat, becomes rapidly volatilized, emitting greenish fumes.

Nitric acid dissolves silver, which thus furnishes several products to medicine and the arts.

The alloys of silver are possible with most of the metals; but, like those of gold, are limited to a certain number of compounds which are employed for the manufacture of articles of luxury.

PLATINUM.

Platinum, or *Platina*, according to recent researches published in Germany, was known by the Romans. Its uses, however, were quite ignored and very few; and it was only in the middle of the last century that, by the exertions of learned manufacturers, it became generally known.

Platinum is grayish-white, and acquires by a polish a brightness, which, however, does not last. This metal is without smell or taste, and possesses tenacity, malleability, and ductility. Its hardness and elasticity are greatly improved by the addition of a very small proportion of iridium. Its specific gravity is 21.50.

Of all the metals, platinum has the smallest dilatation, and is the most difficult to fuse. It becomes soft at a white heat, and in that state may be forged and welded; but its fusion at present can only be effected by the use of the oxyhydrogen blowpipe. This and its high price have prevented this metal from being applied to many industrial uses.

Platinum is dissolved by nitric acid, when alloyed with an excess of silver; it is also dissolved by aqua regia. Caustic alkalies, nitre, alkaline persulphides, phosphorus, arsenic, and chlorine attack it more or less rapidly, with the aid of heat.

The alloys of platinum with most of the metals would certainly be employed, were not its infusibility and its cost a drawback to a general use.

ALUMINIUM.

Aluminum, or *Aluminium*, is of an entirely recent origin; and its employment in the arts dates back only a few years. The industrial development of aluminum is especially due to M. Sainte-Claire-Deville.

Although the uses of this metal have not yet reached

their culminating point, we may foresee that it will be very serviceable.

Already, its manufacture is no longer confined to the limits of the experimental laboratory, its price has considerably decreased, and various trials have shown its usefulness in certain manufactures. The great lightness of aluminium, its malleability, ductility, and difficult oxidation, have retained it for certain uses, but not so many as were expected when it made its appearance in the arts. The specific gravity of aluminium, which does not exceed 2.6, is a characteristic of this metal. Gray, and capable of acquiring a bright, although not lasting polish, aluminium would be more generally employed, were it not so soft, dull in lustre, and expensive.

The chemical properties of aluminium would seem to favor its uses in industry. It is unacted upon by cold nitric and sulphuric acids, by air, water, and steam. Hydrochloric acid dissolves it.

It appears to alloy with many metals, especially with copper, producing certain kinds of bronzes of which we shall speak hereafter, and which already are among the important uses of aluminium.

Generalities, Tables, and Data.

The following tables, borrowed from different authors who have copied from their predecessors, and who could, no more than ourselves, guarantee the accuracy and authenticity of the figures, will terminate all that we have to say concerning the physical and chemical properties of the metals which we have briefly considered.

We advise the reader to consider, as we do, these numbers only as data for relative comparisons, rather than as entirely correct results. This is certainly to be done, when we look at certain points which need the verification of experience, for it would not be possible to admit them without raising certain doubts.

Metals.	Temperature of fusion.	Relative hardness.	Relative ductility through draw plates.	Relative malleability by rolling.	Relative tenacity.	Conductive power for heat.	Conductive power for electricity.	Specific gravity.
	o. c.				kilo.			
Copper.........	1090	6	6	3	137	898	"	8.88
Tin.............	230	10	8	4	16	303	"	7.29
Zinc	410	4	7	7	50	363	"	6.86
Lead	320	11	9	6	12	180	"	11.35
Iron............	1500	2	4	8	250	374	650	7.81
Cast iron	1200	"	"	"	"	"	"	7.21
Bismuth.......	270	9	"	"	"	"	"	9.82
Antimony.....	690	3	"	"	"	"	"	6.71
Nickel..........	"	1	5	9	48	"	"	8.38
Arsenic........	"	"	"	"	"	"	"	5.63
Mercury	Solidification. 39½°—	12	"	"	"	"	100	liquid. 13.60 solid. 14.39
Gold	1000	7	1	1	68	1000	3975	19.36
Silver..........	1000	8	2	2	85	973	5152	10.47
Platinum......	2000	5	3	5	125	981	855	21.50
Aluminium...	760	"	"	"	90	"	"	2.60

Metals.	Specific gravity.	RESISTANCE TO FRACTURE, IN KILOGRAMMES, AND PER SQUARE MILLIMETRE.		COEFFICIENT OF ELASTICITY ACCORDING TO THE	
		Slow.	Sudden.	Longitudinal vibrations.	Extensions.
Lead, cast	11.21	1.25	2.21	1993	1775
Lead, drawn	11.17	2.07	2.36	2278	1803
Lead, annealed	11.23	1.80	2.04	2146	1727.5
Tin, cast	7.40	3.40	4.16	4643	"
Tin, drawn	7.31	2.45	3.00	4006	"
Tin, annealed	7.29	1.70	3.60	4418	"
Gold, drawn	18.51	27.00	27.05	8599	8131.5
Gold, annealed	18.03	10.08	11.00	6372	5584.6
Silver, drawn	10.37	29.00	29.60	7576	7357.7
Silver, annealed	10.30	16.02	16.40	7242	7140.5
Zinc, cast	7.13	1.50	"	7536	"
Zinc, drawn	7.10	12.80	15.77	9555	8734.5
Zinc, annealed	7.06	"	14.00	9272	"
Copper, drawn	8.93	40.30	41.00	12536	12459
Copper, annealed	8.94	30.54	31.60	12540	10519
Platinum, drawn	21.25	34.10	35.00	16159	"
Platinum, annealed	21.20	23.50	26.40	15560	"
Iron, drawn	7.75	61.10	64.00	19903	20869
Iron, annealed	7.76	46.88	50.25	19925	20794
Steel wire	7.72	70.00	87.80	19445	18809
Steel wire, annealed	7.62	40.00	53.90	19200	17278
Nickel, pure	"	90.00	"	"	"
Cobalt	"	115.00	"	"	"
Antimony, cast	6.71	"	0.67	"	"
Bismuth, cast	9.82	"	0.97	"	"

NOTE.—This table and the following one are borrowed from the interesting researches of Mr. Wertheim on the physical properties of alloys. The results given are certainly not free from errors; the notable differences between substances whose analogy is too great for allowing much diversity in their relative resistance and elasticity, is a proof that all the numbers are not sufficiently accurate.

At all events, besides an indication of the specific gravity of alloys which few authors have presented so completely, we find in these tables very interesting data and comparisons for study. When we compare the results of experience with the figures found by calculating, according to the proportion of each metal forming an alloy, we do not find enough regularity to allow us to form a rule by which we may foresee what will be the result of a change in the proportions of the compound. Nevertheless, in practice, we will admit that the coefficient of elasticity may be approximatively deducted from those of the component metals.

Mr. Laboulaye, in his *Dictionnaire des Arts et Manufactures*, where the question of alloys is treated to a certain extent, is astonished because we have not made accurate experiments for arriving, by figures, at the relative value of alloys, as regards their physical qualities, resistance, elasticity, etc. The researches of Mr. Wertheim, though incomplete and insufficient in their results, were made in that direction, and are certainly a progress; but when we come to examine what that work has produced, notwithstanding the conscientious care with which it was done, we must acknowledge that, for some time to come, the alloys will not be studied in the way recommended by Mr. Laboulaye. There are so many unforeseen circumstances, happening even when studying isolated metals, which leave in the dark many important questions, after numberless experiments, that we must not be astonished at the difficulties encountered by those who experiment on alloys.

Alloys.				Specific gravity.	Coefficient of elasticity by vibration.	Maximum of extension.	Cohesion per square millimetre.
							k.
Lead.	68.50	Tin.........	31.50	10.073	2596	0.552	0.93
"	63.80	"	36.20	9.408	2969	2.077	2.46
"	42.50	"	57.40	8.750	3512	1.591	2.07
"	33.25	"	66.75	8.378	3700	0.340	1.07
Lead.	62.40	Bismuth..	37.60	11.037	2021	0.262	1.52
"	50.00	"	50.00	18.790	2367	0.440	1.79
"	33.33	"	66.66	10.403	2838	0.025	5.22
Lead.	76.00	Antimony	24.00	10.101	2183	"	1.87
"	62.00	"	38.00	10.064	2592	"	5.59
"	43.00	"	57.00	8.946	3242	"	"
"	35.00	"	65.00	8.499	3536	"	"
Lead.	95.40	Gold	4.60	11.301	2227	0.055	4.74
Lead.	48.00	Silver	52.00	10.743	3095	"	"
Lead.	98.85	Platinum.	1.15	11.473	2684	0.026	1.65
"	85.00	"	15.00	12.207	3107	"	"
Lead.	95.00	Zinc.......	5.00	11.195	2144	0.069	2.75
"	92.20	"	7.80	11.172	2493	0.060	2.02
"	87.00	"	13.00	11.130	2833	0.060	2.02
"	76.30	"	20.70	0.100	4007	"	2.47
"	68.20	"	31.80	9.043	6647	"	3.40
"	39.00	"	61.00	8.397	6108	"	"
"	24.00	"	76.00	7.910	7352	0.004	4.40
Lead.	94.20	Copper ...	5.80	11.165	2113	0.043	2.13
Tin ...	33.00	Bismuth .	66.00	8.68	3610	0.028	8.19
"	54.60	"	45.40	8.89	2874	0.015	6.63
Tin ...	78.50	Antimony	21.50	7.21	4033	"	8.86
"	66.00	"	44.00	7.05	4695	0.010	7.82
"	67.70	"	42.30	7.007	5168	"	"
Tin ...	78.50	Zinc.......	21 60	7.366	5336	0.246	5.78
"	73.40	"	26.60	7.255	5982	0.252	5.00
"	64.00	"	36.00	7.143	6453	0.036	4.68
"	48.00	"	52.00	7.193	7113	0.124	2.44
"	37.50	"	62.50	6.746	6976	0.082	4.32
"	26.70	"	73.30	6.957	7314	0.023	7.52
Tin ...	96.70	Platinum	3.30	7.578	5309	"	4.75
Tin ...	61.60	Copper....	38.40	8.332	6113	"	"
"	48.30	"	51.70	8.531	8280	"	"
"	21.00	"	79.00	8.813	9784	"	"
"	7.80	"	82.20	8.738	"	"	"
Tin...	98.20	Iron.	1.80	7.266	4881	"	"

Alloys.						Specific gravity.	Coefficient of elasticity by vibration.	Maximum of extension.	Cohesion per square milli metre.
									k.
Silver.	94.50	Copper....	5.50			10.121	8913	"	44.05
"	87.40	"	22.60			9.603	8590	0.002	51.97
Gold..	78.20	Platinum.	21.80			19.650	9844	"	7.12
Gold..	97.25	Iron.......	2.75			18.842	9024	0.016	20.41
Zinc..	76.80	Copper....	23.20			7.301	7678	"	4.10
"	51.50	"	48.50			8.265	8774	"	18.68
"	43 30	"	56.70			8 310	9105	"	36.80
"	33.75	"	65.25			8.606	10163	"	60.20
"	14.60	"	85.40			8.636	9778	0.001	51.90
Lead.	57.00	Antimony	18.00	Tin.	25.00	9.196	2735	0.032	7.80
Lead.	44.50	Bismuth.	47.80	Tin.	17.70	9.795	2626	0.695	1.74
Lead.	73.00	Tin.........	12.00	Zinc.	15.00	10.212	2486	0.162	1.44
Tin ...	51.00	Antimony	28.00	Cop.	21.00	7.751	5770	"	4.17
Zinc..	35.00	Copper....	57.50	Nickel	7.50	8.403	9517	0.001	"
"	18.60	"	60.00	"	21.40	8.541	10227	0.001	61.88
"	37.00	"	43.00	"	20.00	8.436	11722	0.001	55.00
"	21.00	"	50.60	"	8.40	8.615	12250	0.002	68.10

II.

PHYSICAL AND CHEMICAL PROPERTIES OF ALLOYS.

It is to be understood that the following indications must be considered only from a general point of view. When stating the properties acquired by the alloys of metals, we must eliminate those anomalies presented by certain combinations, which are outside of the general limits in which the experimenter works.

Fusibility.—The alloys are generally more fusible than the least fusible of the component metals, and very often more fusible than any of them taken separately. The alloy of Darcet or of Rose, which is a compound of tin, lead, and bismuth, in variable proportions, is a striking example of the principle we have set forth.

Thus, admitting that—

Tin melts at 230° C.;

Lead at 320°;

and Bismuth at 270°; most of the alloys made with these three metals will melt below 100° C. (boiling water).*

However, we must observe that all the alloys do not exactly follow this rule, which is true especially for certain white metals, as those we have named, and to which we must add antimony and arsenic.

Hardness.—Alloys are generally harder and more brittle than the hardest and most brittle of the component metals. Certain soft metals, such as lead, for instance, increase the hardness of the metals with which they are alloyed. Thus, in an alloy of lead and tin, lead may sensibly increase the hardness of tin.

Ductility.—Tenacity.—A few metals, employed singly or united, increase the ductility and tenacity of other metals which are deficient in this respect. However, most alloys have a ductility and tenacity less than that of the most ductile and tenacious of the component metals.

The crystalline structure of alloys has a great influence on their tenacity. Certain alloys, whose crystallization presents large grains, must be very slowly and gradually cooled, if we desire to retain their natural tenacity.

Specific Gravity.—There is no precise law which gives the relation between the specific gravity of an alloy and that of its component metals.

The specific gravity of alloys is sometimes above, sometimes below, that which would be deduced from

* All the temperatures given in this work are according to the Centigrade scale.

the specific gravities, and the proportion of the metals forming the mixture.

The specific gravity of an alloy may be expressed by the formula $\Delta = \frac{(P+p)\,D\,d}{P\,d+p\,D}$, in which P and p are the weights of the metals, and D and d their respective specific gravities.

When there is equality between the result of the formula and the specific gravity of the alloy, there is neither dilatation nor contraction; but if the specific gravity of the alloy, taken by direct experiment, gives a number greater or smaller than Δ, then we arrive at the conclusion that there is contraction or dilatation.

It is hence possible, by the use of the formula, checked by direct experiments, to determine the specific gravity of a certain number of alloys, and to form the following lists of binary alloys which show the graduation of the specific gravity.

I. Alloys, the specific gravity of which is greater than the mean specific gravity of the component metals:—

Copper and zinc.
Copper and tin.
Copper and bismuth.
Copper and antimony.
Lead and antimony.
Lead and bismuth.
Silver and zinc.
Silver and lead.
Silver and tin.
Silver and bismuth.
Silver and antimony.
Gold and zinc.
Gold and tin.
Gold and bismuth.

II. Alloys, the specific gravity of which is less

than the mean specific gravity of the component metals :—

Iron and antimony.
Iron and lead.
Iron and bismuth.
Copper and lead.
Lead and tin.
Tin and antimony.
Zinc and antimony.
Silver and copper.
Gold and silver.
Gold and iron.
Gold and copper.
Gold and lead.

The specific gravity of alloys may give an approximate knowledge of the proportion of the component metals. For instance, we may ascertain the purity of tin by the "trial of the bullet." In a bullet-mould we first cast a ball of pure tin, which will serve as a standard; then we cast in the same mould, alloyed tin, and the greater or less weight of the balls thus obtained indicates a greater or less proportion of lead.

The experiments of Muschenbrœck on the variations of specific gravity of alloys, in which the proportions of the component metals were made to vary, would seem to show that there is a point where the combination is more intimate, and which, very likely, corresponds to alloys in definite proportions.*

We would then be led to admit that the union is more complete, and that there is a tendency to condensation, when the alloy is made of metals having a great affinity for each other. On the other hand, there would be a dilatation when the two metals have little affinity for each other, and are only mixed. Thus copper, which possesses a great affinity for zinc and tin, forms with

* See the preceding tables by Mr. Wertheim.

these metals alloys having a specific gravity greater than the mean.

Elasticity.—Mr. Wertheim, who has closely studied the interesting question of alloys, has tried to ascertain the ratio which exists between the mechanical properties of metals and of alloys, in order to determine the molecular disposition of these compounds.

The alloys were prepared with pure metals; and these were carefully mixed, stirred, and cast in heated moulds.

Each alloy was submitted to chemical analysis; and when, by volatilization, oxidation, or liquation, it departed from the original composition, it was rigorously put aside.

The experiments of Mr. Wertheim were carried on with about sixty binary or tertiary alloys used in the arts. Among them were many well-known alloys, whose mechanical properties had been more or less investigated by several authors; as, for instance, bronze, brass, similor, type metal, bell metal, gong metal, cymbals, etc.

The results, as given by Mr. Wertheim, may be summed up as follows:—

1. Alloys behave like single metals, as regards vibration and expansion.

2. The cohesion, and the limit of elasticity or of expansion, cannot be determined *primâ facie* from the data known for each component metal.

3. The coefficients of elasticity of alloys agree quite well with the average of the coefficients of the component metals. The contractions or dilatations have little influence on these coefficients. We may, therefore, determine beforehand the composition of an alloy, which should have a certain elasticity, or conduct the sound with a given velocity; provided that either of these conditions remains between the extreme limits of the coefficients of each of the component metals.

4. The coefficient of elasticity is greater as the molecular arrangement is closer, and the grain finer and more homogeneous.

Specific Heat.—The researches of Mr. Regnault on specific heat have shown that the average specific heat of the component metals is sensibly that of the alloys; provided that the observations are made at an average temperature sufficiently remote from the points of fusion and of softening.

Latent Heat.—Mr. Rudberg, who has made remarkable researches on the properties of latent heat, has ascertained that, when a melted alloy is allowed to cool, the thermometer becomes generally twice stationary between the points of fusion and solidification. Of these two indications of the thermometer, one is constant for every alloy of the same two metals, and the other varies with their respective proportions.

Two metals melted together, according to Mr. Rudberg, should form a combination in definite proportions, which inclines towards the one in excess. The chemical alloy, when alone, becomes solid at a determined point, which Mr. Rudberg calls the "constant point." But when one of the two metals is in excess, the solidification of the metal and of the alloy does not take place at the same point; the metal in excess, which has a tendency to become solid first, loses its latent heat, and produces a stoppage in the descent of the thermometer. The metal which has first solidified is dispersed through the chemical alloy which remains fluid, and this, in its turn, becoming solid, causes the second stoppage at the thermometer.

Thus, lead becomes solid at 325° C., tin at 228°; and for the alloys of tin, the "constant point," or point of fusion of the chemical alloy, remains at 187°.

Oxidation.—The alloys are not generally so easily oxidized as their component metals, when taken singly. In some cases, however, oxidation is greater in the alloys. That of lead and tin, for instance, when lead is in excess, burns and becomes oxidized very rapidly at a red heat.

When one of the component metals is easily oxidized, and is united in an alloy with another metal which is not, or very little, oxidable, it is possible to separate the metals by transforming the former into an oxide, while the latter remains unchanged. This property is the basis of the operation of cupellation, by which silver is separated from lead. We may, by a similar operation, separate two metals differently oxidable, the more oxidable being much more rapidly oxidized than the other.

The oxidation of alloys, under the influence of atmospheric dampness, is generally less than that of the component metal which is the most easily oxidized. It happens also, in statuary bronze for instance, that the alloy becomes rapidly oxidized at the beginning, more so than would the metals, taken singly and similarly exposed; but after this first effect, the oxidation seems stopped, and is not so destructive as would be the case for the isolated metals.

Although acids appear to act upon the alloys the same as upon the principal metal of the composition, we must also admit that, after a certain length of time, their action will be less destructive for the alloy than for each single metal.

III.

PREPARATION AND COMPOSITION OF ALLOYS.

ALLOYS are made all at once, that is, by combining the metals in the same crucible, or in the same furnace, in one operation.

Or they are made in several operations, that is to say, by uniting first two metals, then three, and so on, in order to obtain a more complete alloy, by the aid of previous combinations already prepared.

By the first method, which is that generally practised, the combination is never so intimate that, notwithstanding the care given to the operations of fusion, stirring, and casting, we may consider the alloy perfectly dense, regular, and homogeneous in all its parts.

We arrive at greater accuracy by the second process. The combinations made separately of metals having a mutual affinity, allow of more precision in the proportions, and more facility in the formation of complex alloys, than would be the case if the metals were added one after the other.

The order in which metals are added to an alloy is far from being a matter of indifference. Indeed, it would not be sufficient, for obtaining a good result, to throw into a crucible, without method or rule or measure, metals whose properties of assimilation are too far apart to combine in a satisfactory manner.

In an alloy of copper, tin, and zinc, for instance, it is preferable to add the tin to the melted copper, and then the zinc, than to introduce the zinc first, and the tin afterwards.

In the quaternary alloy of copper, tin, zinc, and lead, we prefer the order in which the names are here stated. The lead, especially, is to be added the last.

Many other examples could be given in assertion of this rule, which are worth remembering, and are based upon experience and a knowledge of the metals.

By a ready experiment we may ascertain the truth of these principles, and see that the method employed for producing an alloy is not without influence.

Let us combine 10 parts of copper with 90 parts of tin, to which we add 10 parts of antimony. On the other hand, let us combine 10 parts of copper and 10 parts of antimony, to which we add 90 parts of tin.

We have two alloys, which, chemically speaking, are the same; but we may readily ascertain that they are widely different as regards fusibility, tenacity, hardness, etc. These transformations, which appear in combinations the component parts and proportions of which are the same, are evidently due to a peculiar molecular arrangement, produced in the alloy by the order in which the component metals have been added.

In the alloys made in one operation, whatever be the care taken in the fusion and the stirring, the chance is less for the combination to be homogeneous, the greater the difference in the specific gravities of the component metals. When casting, there is a "parting" or liquation by which the heaviest metal goes to the bottom of the mould.

This liquation is to be seen especially in the alloys of copper and tin, which, when the castings are considerable, retain with great difficulty the same homogeneousness and proportions throughout the full extent of the pieces.

The difference of specific gravity is not the only cause which produces the separation in castings at the time of cooling. When an alloy begins to congeal, there is generally formed a less fusible alloy, which becomes solid in proximity to the cooling surfaces, and another more fusible and lighter alloy, which has a tendency to form an upward current in the centre of the piece.

This separation of metals in a fused alloy causes great difficulty in the manufacture of bronze ordnance, where the separation of the tin produces whitish spots, more fusible than the remainder of the metal, and which are melted and removed by the heat of the burning powder.

A rapid and powerful cooling is the only way to

prevent such results, which cause the rapid destruction of ordnance. The separation is prevented entirely or partially if the alloy solidifies as soon as it is placed in the mould.

A slow cooling is always an impediment to the homogeneousness of alloys. When it does not produce a separation of the metals, it occasions a state of crystallization, easily seen, which is always detrimental to the solidity of the metal.

This crystallization will generally increase the hardness of the alloy, but impairs its tenacity considerably. It appears especially in certain alloys which, retaining for a long time a high temperature, when cast are subject to settlings and shrinkages. But this crystallization, and all its accompanying evils, may be prevented by means of large runners, heavy enough to weigh on the metal, and by accessory means which aid in a rapid cooling, such as shaking the moulds after casting, throwing water on certain parts of them, etc.

However, it is a mistake to believe that, in order to obtain a more rapid cooling, it is proper to cast at a low temperature those alloys which have a tendency to crystallize, to shrink, or to lose their shape.

All alloys, as a rule, gain by being cast at the highest temperature proper to each of them, taking care not to increase the loss too much by volatilization or oxidation. An alloy which is cast when hot, cools off in better condition than an alloy which is run into the moulds in a pasty state, and is not subject to those flaws, blow-holes, and shrinkages to be seen in metals the fluidity of which was incomplete.

The processes of "liquation" employed in the operations of metallurgy for extracting certain metals from less fusible ones, may not require so thorough and regular a heating as is necessary for alloys which are to be cast.

On the one hand, our object is only to extract crude

metals, if we may so term them, which are to be melted and worked up again, before they are fit for use in the arts.

On the other hand, we merely require that temperature which is necessary for separating from the alloy one of the combined metals, which melts, leaving the other metal isolated. Thus, for instance, in order to separate silver from copper, we begin by melting the alloy of silver and copper with such a proportion of lead as to have equal parts of copper and lead in the compound. Then, by heating up to a certain point, two alloys are formed: one which is easily fusible, and contains 12 parts of lead for 1 of copper; and another which is less fusible, and contains 12 parts of copper to 1 of lead. The former melts, carrying with it the $\frac{12}{13}$ of the silver, which may be extracted by cupellation.

The alloys, as we have already said at the beginning of this chapter, may be made at once in one operation, or by fractional operations. Binary alloys, having their own characteristics, may be used for forming other compounds, endowed with other properties.

If these alloys are combined with only one new metal, there generally results a new binary alloy, where the first alloy acts like an elementary metal. If the combination takes place between two alloys previously made, there is formed a new compound whose properties may be very different from those of an alloy made by combining successively each metal.

The binary alloys have a real importance in this way, that, with them, the peculiar qualities of both metals may be turned to the greatest account. But these alloys, whether they are wanting in cohesion, or because they do not entirely possess those qualities required in the arts, should be modified by the addition of new metals. These produce a sort of "hybrid" with the former metals of the alloy, and the combinations are quite different from those where the metals were united two by two. At all events, such alloys are more intimate and homogeneous.

In general, it is advantageous to introduce into the alloys a certain number of elements, even in small proportions for many of them, and although several of these elements would not appear to possess an appreciable utility, or have an important effect. The results of affinity obtained by the new elements favor the mixtures, increase the density and the homogeneousness, at the same time that they sometimes counterbalance, with great advantage, the tendency to liquation or separation in the melted mass.

Thus, for instance, a statuary bronze, which could be made entirely of copper and tin, acquires new and indispensable qualities by the addition of zinc and lead, even in small proportions.

As another example, the alloy of copper and zinc, which as such might be suitable for certain uses in the arts, becomes much more valuable for these same uses, and is improved and completed, by the addition of a small proportion of tin or lead.

The more complex an alloy is to be, the more important is it that its preparation should be effected by the union of more simple alloys, previously made. Outside of the considerations which guide the founder as to the order in which the metals should be melted, such as the peculiar conditions of affinity, the similitude in the specific gravities and the points of fusion, it is proper to examine the means and processes by which we add to the final melting those metals whose proportions in the alloy are comparatively small.

These various observations will find their confirmation when, further on, we shall state our researches on the alloys of different metals, and examine the principal alloys in actual use in the arts.

As generally practised, the metals to be combined

are melted by processes and in apparatus which vary according to the quantity of alloys to be cast, or the nature of the metals under treatment.

The metals easily fusible, such as lead, tin, etc., are melted in a ladle, or in wrought or cast iron kettles.

The more refractory metals are melted in crucibles, whose qualities of solidity and resistance to the fire are the more sought for, as the metals have a higher point of fusion, or are more valuable.

For gold, silver, and platinum, we require crucibles of a superior quality, which will not crack, and thus lose in the fire the metals they are intended to receive.

For copper and its alloys, although requiring crucibles as solid and lasting as possible, we look more towards economy, because the work is frequent and regular, and we operate on quantities of less value.

When the mass of metal becomes considerable, whether because many castings are to be made, or because of the heavy weight of the pieces, instead of the crucibles, we operate in reverberatory furnaces, and sometimes in cupolas.

The processes of melting and mixing the metals in a crucible, however simple they appear at first sight, require certain precautions upon which we cannot too strongly insist.

The alloys made in one operation are always very difficult of preparation, when the metals, such as zinc and lead, copper and lead, for instance, possess a sort of "antipathy" in their affinity. It is with much trouble that we obtain, in this way, thoroughly homogeneous castings, presenting the same body and grain of similar alloys which have already passed through a previous fusion.

In order to arrive at the best possible results, without employing the method by separate operations, it is proper, as a general rule, to endeavor to operate according to the following principles:—

1. To charge the crucible, and melt first the least fusible of the component metals.

2. When this metal is in fusion, to heat it up to such a point that it will be enabled, without too great a cooling, to bear the introduction of the other component metals.

3. Once the first charge is in fusion, to introduce the other metals in the order of their difficulty to melt.* Whatever are the proportions of the component metals, and no matter which is the basis of the alloy, it is absolutely necessary that the most refractory metal should be melted first. Its fluidity, indeed, gives the measure of the temperature necessary for finishing the alloy. By charging first a fusible metal, it may volatilize and become oxidized, and the crucible may also break by raising the temperature high enough to receive, without too much cooling, a less fusible metal. At the same time, there will be more waste, and the proportion of the alloy will be sensibly changed.

4. To present at the flame of the furnace the metals which are to be subsequently added, in order to heat them as much as possible, and thus facilitate the change of temperature which takes place when the new metal is added to that or those already melted in the crucible. This practice is especially good when we have to introduce a volatile metal, such as zinc, which, being melted too rapidly, may cause the crucible to break.

5. To stir after the introduction and melting of each component metal; and to cover the crucible, at the same time that the fire is increased more or less, according to the less or greater fusibility of the metal.

6. To cover the alloys rich in zinc with a layer of charcoal-dust. This is not necessary when there is not

* This is a general rule, to be applied in most cases; but there are exceptions. For instance, gold will easily dissolve in melted tin, and platinum in many metals. If platinum were first melted, and zinc for instance added, the temperature necessary to obtain the fusion of platinum would be sufficient to volatilize the zinc.—*Trans.*

in the alloy any metal, such as copper or iron, having a high point of fusion; or when the proportion of zinc added does not require a protracted heating, and the alloy may be poured out immediately. With alloys rich in tin, the charcoal-dust will cause the scorification* of part of this metal; therefore it is preferable to cover the surface of the molten mass with refractory sand or pulverized sandstone.

7. To stir thoroughly the molten alloy just before it is cast, and, if possible, during the pouring out. The stirring is to be done with a stick of white wood, burning without splitting; and not with an iron rod, which has a tendency to produce dry alloys, and may modify the nature of the compounds by adding some iron to the alloy—a small proportion, it is true, but nevertheless appreciable.

8. To carefully clean the crucible after each operation, in order to maintain the accuracy of the mixture, and facilitate the fusion.

Such are the main conditions for obtaining alloys in one operation. If alloys thus prepared give some trouble in obtaining good results, they are very economical, and present the advantage of keeping, as strictly as is allowed by the fusion, the proportions of the mixture.

Moreover, in practice, it is generally acknowledged that a small proportion of an old alloy added to a new one, improves it by giving it the homogeneousness which otherwise would be imparted only by a second fusion.

* The author uses the word "scorification," but we do not think that the term is entirely appropriate. Nevertheless, it is certain that charcoal is not favorable to alloys of tin and copper, and that pure clay crucibles are to be preferred to those of plumbago for such alloys. Metallurgists know that at a certain period of the refining of copper, the metal is carburized and brittle. In order to prevent this carburization, it has been recommended to give a coat of pure clay to the interior of plumbago crucibles.—*Trans.*

In ternary or quaternary alloys made of copper, zinc, tin, and lead, it will always be well, in order to obtain more homogeneousness in the final mixture, to alloy beforehand the more fusible metals, such as zinc, tin, and lead; and to combine this first alloy with the copper, under the best conditions possible. In this way the last combination will possess better qualities than an alloy made in one operation.

However, we repeat it, alloys made by the first direct method, although much more simple and economical, do not answer all the wants of the arts, and do not present the same guarantees as those which have been remelted. For instance, runners from bronze or brass castings of a first fusion, when melted again, and when the primitive proportions were good, present a better grain, and a metal without defects, which is more easily worked than another alloy made directly by one operation.

The pieces cast with alloys made by the direct method—we always mean those in which copper is a component part—are possibly less liable to breakage and shrinkage than if made from old metal; but, on the other hand, the surfaces are not so clean, and the grain is not so close and easily worked. Moreover, such alloys are not very fluid, and do not produce sharp casts. These defects are more to be guarded against in the case of statuary and ornamental bronzes than when pieces of machinery are to be produced.

As a rule, the oftener a metal is melted, the more it loses its previous qualities.

This is exemplified by cast iron, which, after having been melted several times, loses part of its softness and tenacity, and becomes hard and brittle. This happens also to all metals, in a greater or less degree. Copper, repeatedly melted, becomes more finely granular and less tenacious. The same applies to tin, zinc, and lead. However, the last two metals become purer by a second fusion, and are altogether improved; but these qualities will disappear, if remelting occurs too often.

The deterioration which takes place in the nature of metals melted singly is due to new combinations during the remelting, and is entirely caused by the manner in which the operation is conducted.

Oxidation by the fire and the air, and the presence of iron, which it is nearly impossible to remove during the fusion, are the principal causes of the deterioration we mention.

It will be understood that these causes act more powerfully when we operate with remelted alloys, which lose their primitive proportions by the waste which takes place. And if an alloy made by the direct method gives satisfactory results, it will evidently lose its qualities by subsequent meltings. We may, it is true, maintain the alloy within the proportional limits of its composition, by re-establishing, as much by guess as by experience, the proportions modified by the preceding fusions; but, despite the precautions taken, it is with the greatest difficulty that we can bring it again to its primitive condition.

The brass-founders, especially those of Paris, have succeeded in casting quite large pieces from crucibles only. The combinations are more certain, and there is less waste, than by any other methods of fusion, considered more simple, rapid, or even economical.

The furnaces for crucibles, on account of the smaller space they occupy, and their less cost, are better adapted to the majority of founders. We shall not here indicate the principles to be followed in melting in crucibles, because they are to be found in our book on foundries.

The main point, as we have already said, is to melt first the more refractory metals—copper, for instance—then to add to the molten mass the other component metals in the order of their resistance to fusion. When it is time to take the crucible out of the fire, the surface of the metal is cleaned off, and the molten alloy stirred with an iron rod—wood is better, when practicable—the more thoroughly as the metals are more difficult to combine. At last the crucible is rapidly removed, and its contents poured into the moulds, avoiding any unnecessary contact with the air, and all causes tending to cool the metal.

When large pieces are to be cast, the fire is so conducted that each crucible will be ready to furnish at the same time its contingent of molten alloy. All the crucibles are rapidly removed from the furnace, and their contents poured into a common basin, from whence the metal is delivered to the mould.

The least delay in the pouring out of the contents of one or several crucibles, the irregularities impossible to avoid in the fusion, a temperature more or less equal, the difficulty of stirring sufficiently well when the contents of all the crucibles are united, make this mode of operating somewhat difficult. To succeed with it, we require a well-disposed shop, allowing easy and rapid movement, and skillful workmen practised in that kind of work.

A properly constructed and conducted reverberatory furnace, and even a cupola, when the use of the latter is well understood, will be found more appropriate and more easy of management for casting large pieces, and that without more expense, and with more rapidity in the fusion.*

The reverberatory furnaces for the fusion of copper alloys slightly differ from those employed for the fusion of cast iron. However, we prefer the furnaces where the hollow part of the hearth is near the bridge wall.

The fusion of the metal already deposited on the

* Our readers will understand that we here refer especially, and industrially, to the fusion of copper and its alloys.

bed of the reverberatory furnace is conducted with more care than would be necessary for cast iron. The fire should not be so sharp and so frequently increased, its intensity should be more regular, especially when the metal begins to soften and is near the point of fusion.

When the metal is melted, and when the temperature for running off is reached, the working door above the hearth is opened, and the more fusible metals which complete the alloy are rapidly added. The whole molten mass is then stirred with an iron ladle with the greatest care; because upon a good stirring depends the intimate union of the component metals.

The alloys of copper and tin, more than others, require a thorough stirring. The tin has a tendency to strike (rise) to the surface of the castings, when the stirring has not been thoroughly effected under the influence of a somewhat high temperature. Some operators prefer to melt the tin in the casting-ladle, and then throw upon it the copper from the reverberatory furnace, stirring the molten mass all the while.

The alloys of copper and zinc are more easily mixed; however, the damper of the chimney of the reverberatory furnace is to be kept down at least two-fifths while the zinc is being introduced; the fire should also not be too brisk. Indeed, if we always need to maintain a good heat when the alloy is made, it is also proper not to increase the temperature too much, otherwise the waste will increase beyond measure. Moreover, when all the metals are together, and before closing the charging door previous to an additional heating, it is a good precaution to throw on the surface of the molten metal a shovelful of charcoal-dust or of silicious sand.

When the time for casting has come, the tap-hole at the bottom of the hearth is opened with an iron bar, and the metal is received into a casting-ladle, the top of which is covered with ignited charcoal, which keeps up the heat and preserves the surface of the metal from the contact of the air. The temperature of the alloys of copper with tin or with zinc becomes rapidly lowered, and, if perfectly sound castings are desired, no time should be lost to pour the metal into the moulds. All currents of air are also to be guarded against, and all openings tending to produce them should be closed during the time of casting.

Reverberatory furnaces are also employed for fusing scoriæ, workshop waste, and those large pieces which cannot be broken or divided for melting in crucibles. When the operation is to be made with old alloys, it is necessary first to determine their composition, and then to add the proportions of the required metals, such as zinc, tin, lead, &c., necessary to bring the alloy to the desired composition. The introduction of the new metals into the molten bath is effected according to the rules already given.

Cupolas may be successfully employed for recasting copper and its alloys. Although many founders hesitate to use cupolas, we are enabled to affirm that they offer great advantages when large pieces, and even the ordinary bronze or brass castings for machinery, are to be melted.

The essential conditions to obtain with cupolas a well-alloyed metal, producing sound castings at a proper temperature, may be thus summed up:—

1. To employ a dense coke, whose broken fragments are of a volume somewhat smaller than those for the fusion of cast iron.

2. To use a cupola of medium height, whose dimensions in the clear are those of a cylinder having a diameter equal to one-fifth of the height, and one or two tuyeres—one opposite the other—giving the blast under feeble pressure. The cupola must be carefully heated before the introduction of the copper.

3. To make smaller charges than in the case of cast iron. From 100 to 125 kilogrammes are enough for a

cupola whose diameter is 0.50 metre, and height 2.50 metres.

4. To attend carefully to the tuyere, in order to be ready to tap off the metal as soon as the last drops of the last charge fall on the hearth.

5. To pour the copper upon the tin already melted in the casting-ladle.

6. To stir carefully and continuously while the copper is running into the ladle, and the mixture is being effected.

7. To cover the surface of the molten alloy in the casting-ladle with ignited charcoal.

In the alloys, where zinc is a component part, it is proper to melt the zinc in a separate vessel, to pour the molten copper into the casting-ladle, and, after having covered the latter with a *brasque,** to let the zinc into the copper through an opening made in the brasque. This same hole is used for introducing the iron rod or the wooden stick, with which to stir. An operation thus performed, by using all the necessary precautions for obtaining an intimate mixture, without oxidation or volatilization of the more fusible metals; by managing the fusion of the copper so as to make the minimum of waste; by adding to the copper in the cupola a few ingots of bronze or brass, old runners, etc., which prepare the copper to be alloyed, and give it a fluidity which, alone, it would not have—will permit the casting of even thin pieces, in a satisfactory way, more rapidly than by the use of crucibles or reverberatory furnaces, and, at all events, more simply and economically.

The waste from alloys of copper and tin is less than that from alloys of copper and zinc, because the latter metal rapidly volatilizes as soon as it is heated to a point slightly above the temperature of its fusion.

* *Brasque* is sometimes charcoal-dust alone, sometimes charcoal-dust mixed with ashes or clay. In the latter case, it is used as a lining for furnaces.—*Trans.*

When we melt in a crucible the filings, turnings, and scraps of brass, the waste may go as far as from 25 to 30 per cent., and it is difficult to obtain a metal pure enough for casting. It is therefore necessary to make ingots, which are melted again, and produce another waste of from 3 to 5 per cent. In a cupola, these scraps, kept inclosed in old copper pipes, or enveloped in rough boxes made of old sheet copper or brass, do not produce more waste than in a crucible, and the metal is hotter.

For the alloys cast into ingots, it is preferable to employ wide and not very deep ingot-moulds, in order to avoid the separation called liquation. In bronze alloys especially, if the ingots are too thick, the tin has a tendency to strike to the surface. This defect is not very serious when the ingots are to be melted again; on the other hand, it is highly prejudicial when the ingots are to be laminated, or drawn out under the hammer.

The waste in alloys is entirely dependent on the duration of the fusion, and the time during which the metals, once melted, are subjected to the temperature of the furnaces. However, with equal care and supervision during the fusion, the proportion of waste ought to be less with the crucibles than with the reverberatory furnace or the cupola.

With crucibles the waste varies with the greater or less skilfulness of the founder, and, excepting accidents and some special cases, remains between 3 and 6 per cent. In cupolas the waste ranges from 4 to 10 per cent.; and in reverberatory furnaces, from 6 to 15 and even 20 per cent. With the reverberatory furnaces, always very difficult of management when the temperature is to be regulated during the fusion, and an oxidizing flame is to be avoided, the most skilful workman is not always sure of the amount of waste he will produce. Therefore, in the large copper-works, the management of the reverberatory furnaces is not intrusted to any but the best workmen; because it is too easy for a workman little trained, to pass in a few minutes from the limits of an ordinary waste to an unusual one.

We have given, in another work, a practical process for determining the proportion of the component metals of an alloy. We think it should find a place here, and complete the explanations given in this chapter.

When we know, for instance, the nature of the elements of a binary alloy, a calculation may give the proportion of each of these elements by the following rule:—

Take, two by two (in pairs), the three differences between the specific gravity of the alloy and that of each of the two combined metals, then multiply each specific gravity by the difference of the two others, and write the two proportions as follows:—

The greatest product is to the total weight of the compound as each of the two other products is to the weights of the two component substances.

EXAMPLE.—What is the weight of each of the two elements, forming an alloy of copper and tin, whose specific gravity is 8.761, and weight 130 kilogrammes; knowing that the specific gravity of copper is 8.788, and that of tin 7.291?

Take successively the three differences between the specific gravities, and multiply each of these differences by the specific gravity which was not part of the subtraction.

$$8.788-7.291=1.497\times 8.761=13.115217$$
$$8.761-7.291=1.470\times 8.788=12.918360$$
$$8.788-8.761=0.027\times 7.291=0.196857$$

Write the proportions in the manner we have indicated:—

13.115217 : 130 : : 12 918360 : x=128.048
13.115217 : 130 : : 0.196857 : x= 1.951

129.999

The alloy is therefore made of 128.048 parts of copper, and 1.951 part of tin ; the approximation is 0.001. By operating in a similar manner, we could find the proportions of a ternary quaternary, etc., alloy.

As a complement of this method, which will be found useful by founders, we shall explain the practical means for determining the specific gravity of a substance.

If we take water as the unit for specific gravity, and if we weigh the substance first in the air, then in water, we find the specific gravity by this rule :—

The difference of the weight in water is to the weight in the air as 1, or the specific gravity of water, is to x, the specific gravity we desire to know.

But, as it may happen that the substance is lighter than water, we then attach to it another heavier body, so as to weigh it in water. We deduct the weight of the two substances in the water from their weight in the air, then the weight in water from the weight in the air of the additional body, and lastly this second difference from the first, which gives a new difference which is to the weight in the air of the lighter substance as 1, or the specific gravity of water, is to x, the desired specific gravity.

By these processes, founders may readily determine the component parts of an alloy, without having recourse to analysis, with which they are not always familiar.

PART II.

I.

ALLOYS OF THE METALS MOST USED IN THE ARTS.

WE give the name of *industrial metals* to those which are in general use in the arts, that is to say, those which, being no longer confined to the limits of the experimental laboratory, may form the basis of a somewhat extended manufacture.

For this reason, iron, copper, zinc, tin, lead, antimony, bismuth, nickel, arsenic, and mercury are the industrial metals.

It is needless to insist on the importance of the first five metals, which will be the subject of our first study; they are intimately connected with every question of construction; they depend on each other, if we may say so, and all of them are often employed united.

" Concerning these metals, which, however, are much better known than the others, science shows us that many facts are to be observed, and many doubts resolved.

" Many applications which, at the present day, are not thought of, will be found for these metals as soon as practice shall develop the properties already known, and discover new facts.

" Such should be the aim of all attempts at improvement in metallurgic works.

"At the same time that the ordinary routine of the works is attended to, a manager should not lose sight of any new fact or result, without trying to understand it, and ascertain if in it there is not the basis of future improvements.

"The science of metals is essentially one of practice. Experiments, although they will not from the start lay open the unknown, will alone point out the proper direction for future studies.

"It is especially when metals are alloyed together, that practice plays an important part. Most of the results are due, if we may say so, to chance. And if from the scale of data already collected, a skilful chemist may foresee a few results and go in advance of facts, it rarely happens that he is enabled to understand all the phenomena which take place, and to deduce from them positive and regular rules."

These few lines, which we insert here as a preamble, were written fifteen years ago as the heading of a pamphlet on alloys, the success of which was due to the entire lack of similar works on this subject, and possibly to the importance of the experiments and of the stated results.

The first part of this chapter comprises these experiments and their results, relative to the alloys of copper, tin, zinc, and lead. The second part will be devoted to the alloys of iron with the above-named metals. But there our subject will be neither so interesting nor so complete, because, up to the present day, we have not been enabled to bring to satisfactory results the series of studies undertaken at a previous time on this special subject, which has been but slightly elucidated by the authors who have written on alloys.

1. Studies on the Alloys of Copper, Zinc, Tin, and Lead.

Few practical men have investigated the question of the alloys made with the above metals, although they form, without doubt, the most important portion of the metallic combinations employed in the arts. Margraff, Berthier, Levol, Bobierre, Hoffmann, and a few others, may be mentioned as the only experimenters who have given to the public a certain number of peculiar data on certain series of alloys applied to various purposes, such as copper sheathings for ships, bronze for coins, statuary bronze, etc.

Other persons, whether learned or practical men, have more or less confined themselves to those recognized alloys, the proportions of which, up to the present day, are considered as articles of faith.

Thus we know that bronzes in these proportions—copper 88, and tin 12, are very good for pieces having to resist friction; copper 78, tin 22, are proper for bells; that copper 75, zinc 25, make good brass, etc., and the aim has always been to remain within these primitive limits.

It results, however, from the combinations which we have experimented upon, that by varying sensibly the above proportions, we may arrive at as good alloys for the same uses; some being more economical, and others more lasting, better colored, more tenacious, etc.

The publication of these experiments has therefore its utility, and will allow a comparison between the results already known, and the new properties derived from new combinations.

In our researches, we have divided the operation into:—

1. Fixing the proportions of the constituent metals.
2. Fusion.
3. Examination of the product.

The determination of the proportions would have been very complicated, had we tried to make all the combinations possible between metals taken two by two, three by three, etc., the ratio of each change in the proportions being the unit.

We would have had thus to undertake an innumerable series of experiments, without any probable gain, because, in the majority of cases, a difference of one unit in the proportion of one of the component metals will not produce a sensible modification in the alloy. We have, therefore, been obliged to operate between limits sufficiently distant from one another to afford a certainty in the results; and whenever doubt existed, we have experimented on new proportions between these limits taken as landmarks.

The proportions have been calculated so as to have a total weight of 0.250 kilogramme (about $\frac{1}{2}$ pound), which is sufficient to give as good indications as could be expected from larger quantities of alloy.

The metals, after each of them had been weighed, were melted in a crucible, and cast into vertical moulds, so as to produce a square rod or bar, 0.10 metre long (about 4 inches), and 0.01 metre (about $\frac{4}{10}$ inch) for the sides, and a button having a diameter of 0.035 metre (about $1\frac{3}{10}$ inch), and a height of 0.015 metre (about $\frac{6}{10}$ inch).

The observations which follow, result from the examination of the produced alloy, and bear equally on the nature and appearance of both the bar and the button. These observations are sufficient to characterize the essential properties of the compounds, and are followed by accurate researches on their tenacity, malleability, ductility, etc. A more exact determination could be made only by comparative numbers, but the time necessary was not at our command.

The series of experiments which we are about to present is, without doubt, the most important in the practice, and may be thus subdivided:—

1st. Alloys of tin, zinc.
2d. " tin, lead.
3d. " tin, zinc, lead.
4th. " zinc, lead.
5th. " copper, tin.
6th. " copper, zinc.
7th. " copper, lead.
8th. " copper, tin, zinc.
9th. " copper, tin, zinc, lead.

We shall point out only the main characteristics of the alloys of these nine subdivisions, and shall follow our examination with general observations on the whole of the experiments. By thus summing up the principal results, the differences resulting from each of the possible combinations of the four metals will be brought in opposition, and compared.

It is needless to say that the elementary metals introduced into the alloys were obtained as pure and of as good a quality as the trade could afford. In order to refine them, and, at the same time, to divide them into small rods easily cut, each of these metals was melted. After this fusion, their specific gravities were:—

Copper	8.675
Zinc	7.080
Tin	7.250
Lead	11.300

These specific gravities will serve as terms of comparison for those of the alloys, if we happen to find the opportunity of determining not only these specific gravities, but also the numerical values of the resistance, elasticity, etc., of these combinations which we have studied.

1st. Alloys of Tin and Zinc.*

No. 1. *Tin* 30, *zinc* 70.—Texture of a dull white color.†—An average shrinkage.—Breaks easily.—The

* We repeat that all the following data belong to special researches on alloys, and that in no case have we bound ourselves to consult what is known in the ordinary practice, and from works on the subject. As regards the results on a large scale of these alloys actually used in the arts, we can but refer to our work on "foundries," where that question has been treated with all the extension it requires.

† The color of the texture, which is characteristic of every alloy, depends on the nature of the mould and the temperature of the

fracture offers larger and brighter facets than zinc.—The metal is denser at the bottom of the mould.—Dry to the file.—A fine file imparts a bluish polish.—Breaks under the chipping chisel.—Slightly sonorous.—Shows an appearance of crystallization at the surface, with a slight bluish-yellow color.

No. 2. *Tin* 25, *zinc* 75.—Texture of a white color, sliding to blue.—Slight settling or shrinkage of the bar only, the same as No. 1.—Bright fracture with large bluish facets, like those of zinc.—The tin seems to be in larger proportion at the bottom of the button, the same as No. 1.—The surface is covered with a kind of skin rather wrinkled than crystalline, with the colors of the iris, light blue, violet, and golden yellow.

No. 3. *Tin* 50, *zinc* 50.—Texture pallid white.—The surface of the button is very smooth, granular and lamellar at the same time, without any appearance of shrinkage; the edges are somewhat round, and do not show plainly the iridescent colors.—The fracture is bright, and finely granular upon a ground tin-white.—Clogs the file a little.—The alloy is well mixed, tough and malleable, without being soft.

No. 4. *Tin* 70, *zinc* 30.—The texture is white, and somewhat shining.—No settling.—Feebly sonorous.—The surface is granular, dead white, with some spots light yellow.—Difficult to break.—Bears the hammering well.—Easily worked with the chisel, which takes off long chips.—Clogs the file.—The fracture, like that of tin, is without brightness and crystallization.—When polished, is not so bright as tin.—The alloy is

alloy, when cast. We have endeavored to keep these conditions sensibly constant in all the experiments, and to give thus a certain utility to our remarks, which otherwise would not have a decided meaning.

The same rule applies to our observations on the exterior surface, and the shrinkage of the button.

more complete and better mixed than the preceding ones.

No. 5. *Tin* 75, *zinc* 25.—Texture tin-white, but without brightness.—No settling.—Surface granular, and dusted like with bright particles.—The upper surface has a tint changeable from yellow to a reddish-blue.—Clogs the file more than No. 4.—Very malleable, although resisting the hammer and the chisel more than No. 4.—Bends without the cracking sound of tin.

No. 6. *Tin* 10, *zinc* 90.—The bar or rod shows, at the fracture, the characteristics of a zinc rod.—Clogs the file more than zinc, and the fracture is not of so dull a gray.—The bottom of the button is soft, and easily receives the impression of a punch.—As with No. 2, tin appears to have become precipitated, and the metal at the bottom is even softer than pure tin.

No. 7. *Tin* 90, *zinc* 10.—The rod presents the jagged fracture* of tin, and the runner could be separated only by cutting it.—The alloy clogs the file less than pure tin.—The button had settled sensibly in the middle, although the edges were sharp.—The alloy is very malleable, although not so soft under the hammer.

No. 8. *Tin* 1, *zinc* 99.—The fracture is like that of zinc, but the facets are not so large.—The lustre is slightly brighter after filing.—The middle of the bar had settled.—The button had also settled in the middle, and the lower part was soft like No. 6, although not so thick, on account of the small proportion of tin in the alloy.—The soft portions are bluish like lead, and are easily streaked by the nail.

No. 9. *Tin* 99, *zinc* 1.—The fracture is slightly granular, not so dull and jagged as that of tin.—

* In the French, "*Fracture arrachée*" means the fracture of certain metals, difficult to break, on account of their softness or fibrous state when torn asunder, and their fracture appears to be composed of fibres of unequal length, parallel or crooked. "*Jagged fracture*" is the nearest translation we can arrive at.—*Trans.*

When polished, is not so bright. There is more shrinkage on the bar than on the button, and the surface of the latter presents dark iridescent colors.

General Observations.—The alloys where the proportion of zinc is the greatest, present in their fracture a crystallization, whose large facets shine like graphite. A very small proportion of tin added to zinc causes this crystallization. In similar circumstances, the exterior of the castings is covered with a yellow-white moreen (*moiré*).

In thick castings, where zinc predominates, there is a tendency to a separation taking place at the bottom of the mould; and, what is remarkable, this tendency grows greater as the proportion of tin becomes smaller, which is exemplified by the separation being more sensible in No. 8 than in No. 6. We may add, as a singular anomaly, that the tin, which has passed through the zinc and has become precipitated, loses its distinctive qualities, and acquires the softness and the bluish dull color of lead.

The color of the alloy of zinc and tin, whether simply cast or filed, becomes brighter in a direct ratio with the proportion of tin contained in it.

The alloys already rich in tin become granular when the proportion of zinc is increased.

The alloy No. 3 (tin 50, zinc 50) has the fracture of iron, but its color is duller.

The alloy No. 9 (tin 99, zinc 1) has a fracture presenting no longer the jagged appearance of tin, and is dull gray and finely granular.

The specific gravity of the alloys of tin and zinc is in proportion to the mean specific gravity of the two metals; therefore the alloys where tin predominates are more dense.

The waste is greater where zinc is in excess; the tin having been put into the crucible after the fusion of the zinc, we infer that most of the waste comes from the zinc.

The addition of 1 per cent. of tin to zinc is sufficient to impart to the latter metal a greater resistance, without diminishing its hardness.

One per cent. of zinc added to tin impairs the flexibility of the latter, and, what is remarkable, prevents its peculiar crackling noise. These two alloys, when the combination is intimate, present no other sensible changes.

The alloy of tin 50 and zinc 50 is the best as regards stiffness and economy. More zinc would produce an alloy not so well mixed, more crystallized, and brittle; more tin would give a metal clogging the file, and too soft. However, for thin and resisting castings, an alloy of tin 70 and zinc 30 is well adapted. The alloys kept between these figures and the proportion of half and half are very resisting and tenacious. Their malleability increases with the proportion of tin.

The alloy of zinc 1 and tin 99, without impairing the malleability of the latter metal, increases its hardness and tenacity for castings.

The alloys where the maximum of zinc is employed, are useful in foundries only for thick pieces; they are then very economical. Up to the proportions of tin 30 and zinc 70, they remain nearly as brittle as zinc itself. The proportion of tin 25 and zinc 75 produces an alloy not so flexible as tin, and less brittle than zinc, which could be adopted for foundry patterns.

The alloys Nos. 6 and 8 appeared to us more brittle than zinc, in those experiments where tin, passing through the molten mass in the mould, had become precipitated to the bottom. We may infer from this, that a quantity of tin sensibly less than 1 per cent. is sufficient to change the nature of zinc.

The proportions of tin 40 and zinc 60 possess but little malleability.

2D. ALLOYS OF TIN AND LEAD.

No. 1. *Tin* 75, *lead* 25.—Grayish-white fracture, which may be produced by hammering, and the appearance of which is not so jagged as that of pure tin. —Clogs the file more than tin, and less than lead.—Less flexible and more malleable than tin.—Moderately ductile. No settling at the button, and very little at the bar.—After being filed, the lustre is somewhat duller than that of tin.—The bar does not produce a colored streak on paper.

No. 2. *Tin* 25, *lead* 75.—The fracture is more jagged than No. 1; it is more like a metal torn asunder than a broken one.—The fracture looks like that of lead, but is of a brighter white color.—Malleable.—Very easily drawn under the hammer, like lead. Adheres to the file, but not so much as lead.—Forms a distinct colored streak on paper.—The settling takes place especially near the runner, and is scarcely noticeable at the button.—The surface* presents little iridescence. —By filing, the polish is dull.

No. 3. *Tin* 50, *lead* 50.—Broken without difficulty by the hammer, when the bar has been notched one millimetre deep all around by a saw.—Although not so hard as tin under the hammer, it is equally malleable, ductile, and resistant.—As hard under the file as tin, but not so bright after being filed.—The button and the runner present the same amount of settling and the same color as a similar casting of tin.—The rod produces a slightly colored streak on paper.

No. 4. *Tin* 90, *lead* 10.—Fracture not very jagged, like that of No. 1.—After a notch with a saw, as with No. 3., the bar was slightly bent when the runner was broken off by the hammer.—The polish by a file remains sensibly the same as that of tin.—The runner

* The surfaces are such as they come from the mould, without being hammered, cut, or filed.

has scarcely any settling, the button none.—The alloy clogs the file a little more than tin, is softer, but its texture resembles tin in many points.—It does not give a colored streak on paper.

No. 5. *Tin* 10, *lead* 90.—Fracture as jagged as that of No. 2.—As soft as No. 2, but much less than pure lead.—Produces a streak on paper nearly as colored as that of lead.—Clogs the file.—Stiffer than lead and not so flexible.—Receives the impression of the nail, the same as No. 2.

The nail leaves a slight impression on No. 3, and none upon Nos. 1 and 4.

GENERAL OBSERVATIONS.—The alloys of tin and lead are easily made; they generally impart more resistance to the lead, without sensibly impairing the qualities of the tin. It would not be impossible to ascertain the proportion of lead in the alloy, by the behavior of the latter under a chisel, a punch, and by the streak it leaves on paper.

No. 4 (Tin 90, lead 10) does not give a colored streak on paper; No. 1 (tin 75, lead 25), a very slight one. Between these two limits, as for instance with an alloy of tin 85 and lead 15, no streaks are to be seen on the paper, and it is therefore a practical means to ascertain that lead remains in these proportions.

The alloys of tin and lead shrink or settle less than either of these metals taken singly; they are not so fluid when melted, and the castings have not the same sharpness.

Lead, added to tin, increases its malleability and ductility, but diminishes its tenacity. Difficult to break even after several successive bendings, tin becomes more brittle when alloyed with lead; the fracture is then more marked than that of lead, whatever may be the proportions in the alloy, the latter metal being more easily separated than tin, but requiring, however, to be torn asunder.

In the alloy No. 4 (tin 90, lead 10), tin preserves its crackling noise, possibly not to the same degree as when pure, but enough to lead into error persons not fully conversant with the metals. This property of the alloy No. 4, which, however, much resembles pure tin, explains the adulterations to be found sometimes in commercial tin. The tests by the streak on paper, and the crackling noise, both favor the adulteration. From the proportions of the alloy No. 4, upwards, it becomes very difficult, unless by long practice, to ascertain immediately the presence of lead with the tin.

On the contrary, in the alloys of zinc with tin, 1 per cent. of zinc is sufficient to destroy the crackling noise of tin. This property alone may help to recognize the alloy, which may also be determined by other characteristics already indicated.

The alloy No. 1 (tin 75, lead 25) produces no crackling noise on bending. Bent at a square angle, it begins to show a fracture, which increases when the bar is straightened again. This effect does not take place when pure tin is bent for the first time; it is even not noticeable with the alloy No. 4, although this latter is more brittle and its fracture not so crooked and jagged as that of tin.

This fracture will be the best test for distinguishing the alloy No. 4, from pure tin; and when coupled with a lower crackling noise, a certain mark left on paper, a darker texture and a duller polish, there will be sufficient means to prevent error. But all these indications are so slight, that all of them must agree, and a practised eye is necessary to discern them.

If the proportions of the alloy No. 4 are changed, the less lead is added, the more difficult will it be to ascertain the presence of lead. This explains why in the trade so little tin free from lead is to be found, even among that claimed as very pure.

The texture of the alloy, on those parts cast in contact with the air, is another means of recognizing the presence of lead. In those alloys where lead is to be found in certain quantity, the texture is less crystallized, and covered with a pellicle more granular or wrinkled. There is less iridescence, and the lustre is darker and more metallic. Besides these practical data, and without having recourse to analytical processes, the consumer has other means for distinguishing the alloys of tin and lead. These means are derived as those we have already sketched, from the nature of the alloys themselves. For instance, we may determine them by their specific gravity, which is proportional to the mean specific gravity of the two alloyed metals.

We may also recognize the alloys of which lead is an important part, when by contact with the air they become covered with a white dust of oxidized lead.

An alloy of tin and lead with more than 70 per cent. of lead begins to be of inferior quality as a solder. The alloys for solder remain within the limits of tin 30, lead 70, for heavy works; and tin 70, lead 30, for soft solders; so that, in these alloys, 30 per cent. is the smallest proportion for either of the two component metals.

The alloys of tin and lead are advantageous for fusible compositions. The proportion of tin 60 and lead 40 gives a compound fusible at about 70° C. By increasing the proportion of tin, the fusibility of the alloy increases also, which agrees with results already established.

3D. ALLOYS OF TIN, ZINC, AND LEAD.

No. 1. *Tin* 76, *zinc* 12, *lead* 12.—Fracture similar to that of steel, with fine and bright grains.—Tough. —Clogs the file slightly.—No settling either on the bar or the button.—Dull white texture.—The lustre acquired by filing rapidly disappears.—Does not leave a colored streak on paper.—The alloy is thoroughly mixed.

No. 2. *Tin* 12, *zinc* 76, *lead* 12.—Like zinc, the fracture is lamellar and jagged at the same time.—Tough, but much less than the preceding.—After being filed, its color is more blue than No. 1, and is not so easily tarnished.—Slight settling.—The natural surfaces are covered with a very wrinkled pellicle, of a gold-yellow color sliding to violet.—The alloy is not so thoroughly mixed as the preceding.—A small portion of the separated tin and lead, 3 millimetres thick, ends the button.

No. 3. *Tin* 12, *zinc* 12, *lead* 76.—Jagged fracture without lustre, resembling both those of lead and tin. —More easily broken than these two metals.—Less flexible than tin, but softer under the hammer.—Harder than lead alone.—Leaves a colored streak on paper.—The alloy is more completely mixed than No. 2, and there is no separation to be seen on the button.—No settling.—A colored pellicle the same as the preceding.—Has the color of lead after being filed.

No. 4. *Tin* 34, *zinc* 33, *lead* 33.—Fracture duller and not so jagged as that of zinc, which, however, it resembles.—When polished, its color is grayish-blue, without brilliancy, and not so marked as that of lead. —The alloy is well mixed, somewhat soft, but resisting and with little flexibility.—Very little settling.—The surfaces resemble those of cast-tin, are light yellow, without iridescence.—Leaves a slightly colored streak on paper.

No. 5. *Tin* 10, *zinc* 45, *lead* 45.—Fracture resembling that of zinc, with triangular and bright facets, on a dull ground.—Resists fracture like a tough body, although somewhat soft.—Possesses little malleability. —No sensible settling.—The surfaces are much wrinkled, bluish-violet sliding to yellow at the corners.—Leaves a streak on paper nearly as colored as that of No. 3.—The file gives a dull gray polish.

No. 6. *Tin* 45, *zinc* 45, *lead* 10.—Fracture resembling that of iron, dull gray with shining points.—Texture granular and slightly crystallized like pure melted tin. —No settling.—The surface is like that of tin.—Leaves scarcely any colored streak on paper.

No. 7. *Tin* 45, *zinc* 10, *lead* 45.—Like tin, the fracture is dry and jagged.—The alloy is more easily broken than the latter metal.—No settling.—The file gives a dull gray polish.—Very malleable and resisting.—Its flexibility is a great deal less than that of tin and lead.—Its streak does not color the paper as much as No. 5.

GENERAL OBSERVATIONS.—The presence of lead in these alloys imparts to them more body and resistance than is possessed by the alloys of tin and zinc alone. However, they clog the file as much as the latter. The fractures are, generally, more marked than those of the alloys of tin and zinc. The alloy No. 4, where the three metals are in equal proportions, and other alloys presenting slight variations, are malleable, although not very ductile, and may be employed with great economy in many cases.

The alloy No. 2, as hard and brittle as zinc, although more resisting, may be successfully employed by founders. Like No. 3, it is cheap, and both will be found more serviceable in foundries than either of the three metals taken singly.

These ternary alloys, which are more thoroughly mixed, and more complete than the alloys of zinc and tin or zinc and lead, present the advantage of being more tough without being more expensive.—Numbers 1, 3, and 7 appear to stand friction very well.—Nos. 2, 4, and 5 will do for pieces requiring more resistance than pure zinc.—No. 6 will do for thin castings requiring a certain malleability. It will also be found serviceable for ornaments, and will bear engraving and chasing. For these uses, Nos. 2, 4, and 5 would be too brittle; and Nos. 1, 2, and 7 too soft and yielding.

All these alloys, when polished, have little lustre, and become rapidly tarnished by exposure or friction. They are not to be used as white metals. But, besides the advantages they offer in foundries, several of them might be applied to the manufacture of types, and in galvanizing metals, etc.

4TH. ALLOYS OF ZINC AND LEAD.

No. 1. *Zinc* 75, *lead* 25.—Same fracture as zinc, a little closer.—The fracture at the lower part of the bar is more finely granular than No. 4; the facets are shining like those of a large grain iron.—The lead has precipitated to the bottom of the button, occupying half of it; the separation is also seen on more than one-sixth of the length of the bar.—The portion of the bar where zinc predominates clogs the file more than pure zinc.—No settling at the surface of the button, which is pale yellow.—A slight settling is to be seen on the bar, near the runner.

No. 2. *Zinc* 25, *lead* 75.—The whole bar presents the characteristics of lead; the runner alone has the appearance and the fracture of zinc. A little below the runner liquation has taken place, and the lead has been precipitated to the bottom, leaving at its junction with the zinc an empty space, like a blown hole.—The lead has also separated in the button, the surface of which is very irregular.—The bar has settled like tin.

No. 3. *Zinc* 50, *lead* 50.—The fracture near the runner is like that of zinc melted several times.—The lead has become separated both in the bar and the button, and occupies one-third of the bar and two-thirds of the button.—No settling on the button.—On the bar, the settling is like that of No. 2.

No. 4. *Zinc* 90, *lead* 10.—The fracture is like that of a finely granular zinc.—The entire bar presents this character, without any separation.—The bar, however, leaves a colored streak on paper, like lead.—Nearly all the lead of the alloy is found precipitated in the button.—The button has settled slightly, and when broken, presents large facets with a few jagged portions, where the zinc is. The presence in the button of nearly the whole of the lead employed for the alloy cannot be well accounted for by the lead having precipitated to the bottom of the crucible, notwithstanding the stirring, because the alloy remaining liquid in the mould for some time, lead would have been able to penetrate part of the bar. The latter, however, contained some lead intimately mixed in the whole mass.

No. 5. *Zinc* 10, *lead* 90.—The fracture is like that of lead, that is to say, appearing more like being torn asunder than a true fracture, and its color is not so dull as that of lead alone.—The bar yields to a punch, the same as lead; however, when filed, it produces a certain noise, presents more resistance to the tool, and the file dust is easily detached; in a word, it is tougher than lead alone.—The button, as in the other examples, contains the lead at the bottom and the alloy of zinc at the surface.—The bar and the button present a settling similar to that of lead.

General Observations.—The five preceding alloys, like all the intermediate compositions which we have tried, were all cast at the same temperature, gradually raised. The alloys were carefully stirred, before taking the crucible off the fire, and while running into the mould. The moulds were of green sand, and so disposed as to be cooled rapidly. Notwithstanding all these precautions, it has not been possible to prevent the separation of the lead, which took place as soon as the alloys were run into the moulds. All the samples present this separation, more or less, according to the proportion of lead in the alloy. We may then infer that the alloys of zinc and lead are not practicable; and that, not alone on account of the difference between the specific gravities of the two metals. Indeed, if this separation of the lead may be due to the specific gravity of this metal, we may also suppose, and with as much appearance of truth, that it is occasioned by the zinc, which alloys nearly as badly with tin, the specific gravity of which is not very different; while, on the other hand, it alloys very well with copper, which has greater specific gravity. This is an anomaly very interesting to observers, and which might preferably be attributed to the difference of the melting points.

However, notwithstanding the separation or liquation, it is certain that a very small proportion of the lead remains united with the zinc, sufficient to modify the nature of the former. Thus, from these alloys, it results that the bars of zinc, slightly impregnated with lead, acquire a great power of resistance under the hammer, become harder, more malleable, and adhere more to the file. They leave a colored streak on paper, which is a proof of the presence of lead, and that an alloy takes place with a very small proportion of the latter metal. This can be verified by the results of No. 4, which presents at the fracture the characteristics of zinc, and of which the properties have been modified.

Those portions of the bars where the separation has taken place, and when the lead predominates in the alloy, present an empty blown place, showing how complete and sudden was the liquation. With No. 2 the lead was scarcely welded to the zinc, although the alloy had been run very hot into the mould.

In all these alloys, when the buttons are broken, the zinc is perfectly distinct from the lead; the two metals appear as if placed one on top of the other, although united, and when the surfaces are smoothed or polished, the line of demarcation is perfectly visible. This peculiar arrangement, more curious than useful, might find an application in a case, where it would be desirable to obtain a casting composed of zinc one way, and of lead, the other.

The fracture of zinc holding a minute quantity of lead is not so bright as that of pure zinc; the crystallization presents smaller facets extending in every direction, instead of being vertical to the plane of fracture, as is the case with pure zinc. Such an alloy, made on a large scale, would not show the nature of the zinc sensibly modified. Only those alloys of the two metals holding a small proportion of zinc or lead, about 1 per cent. for instance, will give good castings, if they are carefully stirred, run into the moulds at a good temperature, and rapidly cooled. The alloys of half and half would be very difficult to produce, if not impossible in practice. A piece of ornamentation, presenting a large surface, and cast horizontally with an alloy of zinc 70 and lead 30, had all its lower portions overcharged with lead separated from the zinc, and the line of separation was full of blown holes. Where the lead predominated, the casting was heavy, without sharpness, more like a paste, and presenting the marks of many bubbles of air which could not escape.

To sum up, in the alloys of zinc and lead, where one of the metals is in small proportion, the other predominating metal is improved. Thus, with No. 4, the zinc has lost part of its brittleness, and adheres more to the file; with No. 5, the lead, naturally soft, has acquired a certain hardness and tenacity, at the same time that it has become less flexible.

As regards the zinc, and the same as with the preceding ternary alloys of zinc, tin, and lead, a small proportion of the latter metal improves the alloy; in a large proportion, there is no alloy, or the product is inferior.

5TH. ALLOYS OF COPPER AND TIN.

No. 1. *Copper* 99, *tin* 1.—Texture of a light violet color.—The polish is light red, without much lustre.—

Granular fracture, spotted with light red or salmon red bubbles.—Soft under the hammer, but does not clog the file as much as pure copper.—Has more tenacity than the latter metal.—The surface of the button is convex, reddish on the edges, and covered in the middle with a scoriated pellicle, like pure red copper.

No. 2. *Copper* 95, *tin* 5.—Texture of a very light violet copper.—The polish is yellow, tending to a pale red.—Granular fracture, somewhat jagged, and of a yellowish-orange color.—No settling on the surface of the button, which is wrinkled like bronze (copper 88, tin 12), with some spots of a brown red color resembling that of pure copper.—It is dryer to the file, harder under the hammer, and more resisting than the preceding alloy.

No. 3. *Copper* 90, *tin* 10.—The texture is dull yellow sliding to a very light violet.—The polish is more of a pale yellow and less reddish than No. 2.—Granular and jagged fracture, of a pale yellow, tending to a whitish-yellow.—The surface of the button presents a small and regular settling, and is covered with a wrinkled and tubercled skin, like that of bronze.—Tough, resisting, standing the hammer well, and somewhat harder than the preceding.

No. 4. *Copper* 80, *tin* 20.—Yellowish-gray texture. —The polish is light yellow, tending to the pale gold-yellow of alloy No. 7, of copper with zinc.—The fracture offers some jagged points, but the remainder is lamellar, with scarcely any grains.—A slight settling on the middle of the surface of the button, which is grayish-white on the edges, and covered on the centre with a grayish-black and granular skin.—More difficult to file, yielding less to the punch, more brittle, and consequently more easy to break than the preceding.

No. 5. *Copper* 75, *tin* 25.—Dull gray texture.—Polish, pale yellow passing to white.—Perfectly smooth fracture, without any granular and jagged appearance, and with a yellowish-white lustre.—A slight settling on the surface of the button, which is nearly smooth, and of a dull grayish-black color.—May be easily filed, although much harder than the preceding.—A punch leaves no mark on the alloy, which breaks under the shock.—It flies under the chisel.

No. 6. *Copper* 65, *tin* 35.—Grayish-white texture, with more glitter than the preceding alloys.—Grayish-white lustre, intermediate between iron-white and silver-white.—The fracture is not jagged, although not so smooth and clean as the preceding; it is whiter and has more lustre.—Breaks easily to splinters; cannot be chiselled; very hard to file, and receives no mark from a punch.

No. 7. *Copper* 50, *tin* 50.—Grayish-white texture, not very brilliant, and tending more to white than to gray.—Lustre, grayish-white with a dull reflection.—The ratio between the lustre of the texture and that of the fracture is more direct than in the preceding alloys, where the ratio is inverse. Fracture, white like that of No. 6, but with less lustre.—As brittle and easily broken as No. 6, it is not so difficult to file, but cannot be chiselled.—The surface of the button is smooth, of a dirty yellowish-gray color, and covered with a whitish dust, like the alloys of copper and zinc.

No. 8. *Copper* 40, *tin* 60.—Texture like that of No. 7.—When polished, the lustre is white, with a dull reflection like the preceding, but is much more easily filed and polished.—Between this and No. 7, the difference of action of the file is very considerable; No. 7 is scarcely attacked by the file, while this alloy may be filed nearly like lead, with this difference, that the filings are dryer, finer, and do not clog the file.—The surface of the button is smooth like that of No. 7, and also covered with a dust of oxide of tin.

No. 9. *Copper* 30, *tin* 70.—Texture like Nos. 7 and 8.—Is filed and polished like the preceding, to which it bears much resemblance.—Easily receives the mark of a punch or hammer, although very brittle.—The fracture presents large laminæ, with a lustre like that of No. 8.—The fracture of Nos. 7 and 8 was not lamellar, although not so smooth as No. 6; it was characterized by a few hollow spots, as if stamped.

No. 10. *Copper* 20, *tin* 80.—Texture like Nos. 8 and 9.—The same characteristics of these two numbers.—The surface of the button is smooth, with a few grayish-black crevices.—Receives the mark of the punch well.

Nos. 11 and 12. *Copper* 10, *tin* 90; *Copper* 5, *tin* 95. —The fracture becomes granular and loses its lustre. —Their texture is of a more grayish-white than the four preceding alloys, and they are much less brittle. —They are easily filed, although they hang more to the file, and produce coarser filings. Their polish is whiter, with more brilliancy.

No. 13. *Copper* 1, *tin* 99.—Grayish-white texture, without the brilliancy of that of tin.—The fracture is bright.—Not so easily broken as Nos. 11, and 12, although without much tenacity.—Is easily filed, and chiselled with difficulty, although more yielding than the preceding alloys.

General Observations.—These thirteen alloys are sufficient to give an idea of the anomalies presented by tin alloyed with copper. In the alloys where copper predominates, up to the combination of 85 copper and about 15 tin, the metals obtained are tough, tenacious, with a certain malleability, receiving a fine polish, and very useful in the arts. From the proportion of 15 per cent. of tin, the alloys become harder, dryer, more brittle and difficult to file, until the proportion is copper 75 and tin 25.—The alloy of copper 65 and tin 35 is very brittle, with a fracture like that of white pig-iron, and is scarcely attacked by the file. This brittleness and hardness remain up to the proportions of half and half. However, the alloy of copper 50 and tin 50 is more easily filed, and the other alloys, where tin predominates, reacquire that property which they had lost between the alloys No. 4, and No. 7. The combinations 11, 12, and 13 recover a certain tenacity, become softer, not so brittle, and may be more serviceable, whether as anti-friction metals, or white metals.

The worst alloys, therefore, are not those where tin largely predominates, as is generally believed, and as we have ourselves stated in our book on the foundry. The less useful series, on account of their excess of brittleness and hardness, are, according to our experiments, those limited between the proportions of copper 85, tin 15, and copper 20, tin 80. We must except, however, the sonorous alloys, which reach their maximum of sonorousness with proportions of about 75 of copper and 25 of tin, corresponding to the alloys for gongs and cymbals. The bell metal varies between copper 79, tin 21, and copper 77, tin 23. These alloys, as we have seen, are filed with great difficulty, and the results of our experiments agree entirely with those of ordinary practice. We must also notice, among the alloys which we have pointed to as of little service in the requirements of industry, the alloy No. 6, or one not very different, which is employed as speculum metal for telescopes. The perfectly white color of this metal adapts it to that particular use.

In the alloys ranging from No. 1 to No. 4, a change in the proportions of tin gives various metals with properties sensibly modified.

The composition No. 1 is that of a bronze for medals and coin; it is the only one which is sufficiently malleable when cold to make it worth while to notice this property. The malleability, at the ordinary temperature, disappears with the compound No. 2, but will remain at a cherry-red heat up to the proportion of copper 85 and tin 15. The combinations remaining between No. 3 (copper 90, tin 10) and No. 4 (copper 80, tin 20), comprise the bronzes for machinery. For a red bronze, we adopt the proportions of No. 3; for an ordinary bronze, having a fine orange-yellow color, tough, tenacious, and bearing friction well, without being too hard, we prefer the proportions of copper 88 and tin 12. But copper 85 and tin 15 will give the maximum of hardness and resistance, and the alloy may be filed.

The alloys ranging from No. 2 to No. 4, where the proportion of tin is comparatively small, are difficult to produce by a direct operation. The mixture is often incomplete, and, whatever is the care given to the stirring, the tin has always a tendency to strike to the surface of the castings, and to become thus separated from the copper. We have indicated in our work on the "foundry," the best means for preventing that defect, and producing sound alloys by a direct operation. The manufacture of bronzes for machinery is sometimes conducted on a large scale, and we have given directions for the use of the cupola. Without repeating here what is already known, we shall however state as an important fact, that, when bronze is melted in a cupola, where a few fusions of cast iron have been previously made, its quality is sensibly improved. This result, which is due to the alloy of a small proportion of iron with the bronze, will be noticed when speaking of the alloys of iron with other metals. Therefore, for the sake of economy and to improve the quality, it is preferable to employ a cupola which has already been used, when we desire to melt large quantities of bronze. Pure copper, when melted in a new cupola, wastes a great deal and penetrates the lining of the hearth, when the temperature is raised too much; while this defect will not take place in an old furnace, the lining of which has become hard and vitrified by previous fusions.

The unfavorable results presented by new cupolas are not confined to bronze alone; every founder knows that cast iron becomes hard and brittle at the first melting in a new cupola.

The alloys of copper and tin, where the proportion of the latter metal predominates, are very apt to become oxidized. Generally, the oxidation of tin begins to be noticeable when the proportions reach two parts of copper to one of tin.

That tin will have a tendency to separate from the copper, and strike to the surface of the casting, is not the only annoyance to be feared; we have also to provide against the penetration of the metal into the material of the moulds, and its combination with the sand. When this happens, there is not only danger for the success of the casting, but the waste increases, and the quality of the alloy is sensibly impaired. The facility with which tin separates from the copper and infiltrates the sand of the moulds, cannot be opposed except by an intimate mixture of the two metals, a thorough stirring, running in at a good temperature, and the employment of moulding sands sufficiently wet. Sands, whether too wet or too dry, have an equal tendency to become saturated with the tin which separates from the alloy. It is obvious that this separation of tin is to be feared only in large pieces, when the cooling is slow, and the alloy remains liquid for a long time.

A sample of sand thus impregnated with metal, after the casting of a large journal box composed of copper 88 and tin 12, had a specific gravity of 4.456, while those of the casting and of the pure sand were respectively 7.538 and 1.225. We have thought it useful to notice this fact, although foreign to the results of our experiments.

6TH. ALLOYS OF COPPER AND ZINC.

No. 1. *Copper* 99, *zinc* 1.—Violet texture.—Polish pale red.—Fracture jagged and brighter than that of pure copper, although lighter colored.—More difficult to break than the latter.—Somewhat harder under the file.—The surface of the button is scorified and puffed up, like that of pure copper.

No. 2. *Copper* 95, *zinc* 5.—Violet texture, similar to No. 1.—The polish is a very pale red, tending to yellow.—Fracture tough, jagged, and of a red color, passing to yellow.—Malleable, and difficult to break, even after having been bent several times.—A little harder to the file than the preceding.—The surface of the button is bloated, wavy, but not so scorified as No. 1.

No. 3. *Copper* 90, *zinc* 10.—The texture is neither so violet nor so dark as the two preceding alloys.—The polish is yellowish-red, tending more to yellow.—Fracture finely granular, and yellowish-red.—Not very difficult to break after having been notched with a file.—Bears the hammer well.—Harder to the file than No. 2.—The surface of the button is puffed up on the edges, and slightly settled in the middle; it is covered with a brown skin with reddish-violet spots.—This surface differs more from that of pure copper than the buttons of No. 1 and No. 2.

No. 4. *Copper* 80, *zinc* 20.—Texture violet, sliding to dull gray.—Polish dark yellow without red reflection.—The fracture is more coarsely granular than the preceding one, and of a yellow color resembling gold-yellow.—More difficult to break than the preceding.—Very malleable.—Harder to file than No. 3.—The surface of the button has settled in the middle, and has its edges rounded; its skin is somewhat wrinkled, dark yellow, and presents violet spots as No. 3.

No. 5. *Copper* 75, *zinc* 25.—The texture is light violet, with yellow marbled veins.—Polish dark gold-yellow.—The fracture is finely granular, and the gold-yellow color will rarely appear, unless by filing.—The surface of the button is smooth, slightly granular, without settling, not so dark yellow as the preceding, and with very few violet spots.

No. 6. *Copper* 65, *zinc* 35.—A light yellowish-green texture.—Polish yellow, with a greenish reflection, and brighter than No. 5.—Fracture of a yellowish-orange color, and the facets converge towards the centre.—More easily broken than the preceding, and does not hang to the file so much.—The surface of the button is bloated and dirty yellow, with a few spots of a brighter yellow.

No. 7. *Copper* 50, *zinc* 50.—Yellow texture, sliding to a dull gray.—Polish a pale yellowish-red, as the bronzes of copper and tin.—Fracture dark gold-yellow, with large facets presenting a jagged appearance.—Harder to file than the preceding, and slides under the tool.—The surface of the button is scorified, and grayish-yellow, with a few brighter spots.

No. 8. *Copper* 40, *zinc* 60.—Dirty and dull yellow texture.—Polish yellow, tending to white.—Very hard to file.—Very brittle.—The fracture is smooth, without any grains or facets, like that of a very white pig-iron.—This fracture is very bright, and more so than the polish of the filed metal; its brilliant white appearance imitates that of silver.—The surface of the button is slightly settled and scorified, and is spotted with bright yellow spangles.—The fracture of this button, effected while the metal was quite hot, is as smooth as that of the bar, and with a brilliant lustre, resembling more that of gold than that of silver.

No. 9. *Copper* 30, *zinc* 70.—Texture, a dirty gray, without any lustre.—Fracture smooth, but not so even as that of No. 8; the lustre is not so sensible as with No. 8, although considerable.—Very difficult to file.—Very brittle.—The surface of the button is settled in the middle and covered with a dull grayish-black skin.—This experiment has been made twice; the first sample presented a duller and more granular fracture, of a white color passing to blue and violet.

No. 10. *Copper* 20, *zinc* 80.—Texture, a very dark grayish-black.—Polish dull grayish-white.—Granular fracture, the tint being grayish-white, with a few bright spots.—Very brittle.—Very hard to file.—May be reduced to powder by hammering, the same as the two preceding numbers.—However, a punch will leave its mark on it better than on these two numbers, which do not stand the pressure at all, and fall to pieces immediately.—The surface of the button is swollen, and covered with a bloated and gray skin, without lustre, like the texture, and with a few whitish spots of oxidized zinc.

No. 11. *Copper* 10, *zinc* 90.—Texture dull gray, sliding less to black than the preceding.—More easily filed.—A great deal less brittle.—The polish has not much lustre, and is white tending to gray.—The fracture on a lead-white ground is half granular and half lamellar, with facets having a certain brightness.—The button does not show any sensible settling, and is covered with a very wrinkled, blackish skin.

No. 12. *Copper* 5, *zinc* 95.—Texture a duller gray than the preceding number.—Harder to file than zinc, but softer than Nos. 8, 9, and 10.—Polish dull, with gray reflection.—Fracture a grayish-blue, with bright facets, which are similar to those of zinc.—The surface of the button is smooth, presenting a general shrinkage or settling, and has a dull light-gray color.

No. 13. *Copper* 1, *zinc* 99.—The texture is not so bright as that of zinc, and the gray color is more saddened.—It possesses less lustre, and a white color sliding to a dark one, more than zinc.—Fracture imitating that of zinc, but the ground is darker and the grains finer.—The surface of the button presents the same color as the preceding, but there is more shrinkage.—This alloy is harder, more resisting, and more difficult to file than zinc.

GENERAL OBSERVATIONS.—The series of the alloys of copper and zinc, like that of the alloys of copper and tin, presents the same general analogies in the nature of the compounds, according as one or the other metal predominates in the compound.

The malleability, ductility, smoothness, and firmness of the grain seem to increase with the proportion of copper, to disappear when the two metals are nearly in equal proportions, and to reappear, to a certain degree, when zinc predominates.

Up to No. 7, where the proportions of the two metals are equal, the alloys of copper and zinc are in general use in the arts. With a small amount of zinc, as in all the alloys comprised between Nos. 1, 2, 3, and 4, the products are tough, tenacious, very malleable and ductile, but the objection to them is that they are somewhat expensive. This, evidently, is the only reason why they are but little employed; and manufacturers will even prefer the alloys of copper and tin, made in the same proportions, although more costly, because they are harder, more resisting, more sonorous, and bearing friction better, which qualities are to be found to a less degree in the corresponding alloys of copper and zinc.

The next compounds, comprised between Nos. 4 and 6, are those most used in the arts.

The alloys of copper and zinc, known under the name of brass, and used for pieces of machinery, are generally composed of copper 75 and zinc 25, corresponding to No. 6. Questions of economy will decide whether the quantity of zinc is to be above or below this proportion.

No. 7, where the combination was difficult because of the considerable waste of zinc, had the appearance of a tin bronze, judging by the texture, and the polish after filing. A not over-scrupulous founder, having to deal with a consumer not very well conversant in alloys, may pass No. 7 as a bronze; but if by its external appearance this alloy looks like a bronze, it is easy to ascertain that it is wanting in hardness, cohesion, and even in color, because its polish rapidly becomes tarnished. A little lead added to this alloy gives it more body, and may render it very useful and economical for those castings requiring no chasing, and having no strains to bear.

The compounds Nos. 8, 9, 10 comprise the series of alloys of copper and zinc that are the least serviceable, and are the most brittle, and the dryest, and hardest under the file or the hammer. No. 8, especially, is very brittle, and will fall to pieces by the slightest shock.

If No. 11 begins to acquire a certain firmness, it nevertheless remains very brittle, and of a dull appearance. We do not believe it more serviceable than the three preceding numbers.

Nos. 12 and 13 possess properties similar to those of zinc; they are harder and tougher than the latter metal, and this explains why they are sometimes used, especially those economical combinations approaching that of No. 13.

The direct combination of the alloys of copper and zinc is the more difficult as the proportion of zinc is more considerable. From No. 5 upwards, unless great precautions are taken, a considerable proportion of zinc volatilizes. If, however, care is taken not to keep the copper melted at too high a temperature, to add the zinc in several portions instead of all at once, to heat the zinc previously nearly to its point of fusion, to keep the crucible covered, to have a moderate fire until the moment has come for casting, and then to stir and operate rapidly, we avoid much waste of zinc, and the alloy may be produced in the desired proportions.

At all events, the alloys of copper and zinc, once the proportion of zinc is above 50 per cent., do not seem to us to be worth more extended studies than those already indicated.

By adding another metal, lead, for instance, we thought that we could arrive at better results; but new experiments with lead in small proportion, gave us samples of alloys sensibly the same as those of copper and zinc.

In the metal works, where brass sheets and wires are manufactured, the alloys of copper and zinc in various proportions have, of course, been experimented upon. The results did not correspond to the requirements, when the proportion of copper was less than in No. 6. Altogether, these experiments agree with our own.

We shall notice several of them as a comparison.

Copper.	Zinc.	
30	70	Brittle alloy, with a gray and lamellar fracture like that of zinc.
35	65	Dry, and more brittle than glass:—Conchoidal fracture with the brilliancy of silver.
40	60	Same brittleness and lustre, with a slight yellow tint.
45	55	Brittle, and reddish-gray, but purplish at the fracture.
50	50	But little tenacity, breaking with a jagged fracture of a fine gold-yellow.—Very hard to file; the tool removing this fine color.
55	45	More tenacious and resisting than the preceding alloy; the striæ of the fracture become flat and lamellar, some being yellow and others reddish.
60	40	Resisting. It was necessary to notch it with a chisel, before it could be broken. The laminæ at the fracture are flat and grayish-yellow.

These alloys confirm what we have stated in principle, that the more useful combinations remain between Nos. 4 and 6. We must remark, however, that between Nos. 3 and 4 are to be found the alloys known in the trade under the names of *similor*, *pinsbeck* or *pinchbeck*, *Prince Robert's metal*, &c. The more important of these compounds are:—

Copper.	Zinc.	
80	20	Shining fracture of a fine yellow color.
84	16	Of a finer yellow than the preceding.
86	14	More yellow, and more brilliant.
88	12	A gold-color, and finer grained.

With a smaller proportion of zinc, the alloys are improved; but then arsenic is added to them, in order to make the white coppers; tin, for the manufacture of *chrysocal*; tin and lead, for bronzes for statuary and gilding, &c. &c.

We shall examine all of these compounds further on.

7th. Alloys of Copper and Lead.

No. 1. *Copper* 99, *lead* 1.—Texture, reddish-violet, like pure copper.—Polish, more pallid than pure copper.—The fracture is not so jagged as that of pure copper, and therefore more easily effected; its appearance is dull, with whitish or pink mottled laminæ, more pallid and with a more mixed coloration than is the case with pure copper.—Under the file, acts like copper, although more yielding to the hammer.—The surface of the button is dull black, bloated, and settled like copper.

No. 2. *Copper* 90, *lead* 10.—Texture, light violet, sliding to yellow.—The polish is not as bright as the preceding, and its color is a light pink.—The fracture on a pink ground, mixed with gray on the edges, presents laminæ converging towards the centre.—The button is smooth, with a slight settling, and is covered with a grayish-black pellicle having a certain lustre.—The metal of the button is granular, gray mixed with pink, and more brittle than that of the bar.—The bar clogs the file, and is softer under the hammer than the preceding alloy.

No. 3. *Copper* 75, *lead* 25.—Texture, gray, slightly pinkish.—Polish, without much lustre, and a light pink sliding to gray.—Fracture, light pink mottled with gray, and with closer laminæ than the preceding alloy.—Does not break so easily.—The button is similar to that of No. 2, but its texture is more pallid.—The bar clogs the file, and its resistance to the hammer equals that of No. 2.

No. 4. *Copper* 50, *lead* 50.—Its texture and polish are the same as the preceding.—The colors of the fracture are more mixed, and it is more granular than lamellar.—In this sample, as with the preceding, the lead, penetrating the sides of the mould, has become deposited on the surface of the bar, which is covered with a pink-gray film.—The polished surface shows different tints, tending from light red to gray.—Yields to the hammer, and clogs the file, the same as No. 3.

No. 5. *Copper* 25, *lead* 75.—Presents the general characteristics of lead, although not so yielding to the hammer.—Is more brittle, and breaks with a somewhat granular fracture, without a jagged appearance.

No. 6. *Copper* 10, *lead* 90.—Similar to the preceding, that is to say, more brittle, less malleable than lead, and with a fracture not so jagged.

No. 7. *Copper* 1, *lead* 99.—Similar to Nos. 5 and 6. —The presence of the small proportion of copper is scarcely perceptible, except by a few yellow tints on the surface of the bar.

General Observations.—The alloys of copper and lead are difficult to produce in extreme limits. They are, however, more easy when copper predominates. When the proportion of lead is in excess, this metal cools off the copper in the crucible, or becomes partly oxidized, if the temperature is increased in order to obtain a more thorough mixture. On the other hand, the copper has a tendency to strike to the surface, when the alloy is run into the moulds very hot. It results then, that the alloys of copper and lead are difficult to obtain by the direct process, in one fusion.

The alloys Nos. 3 and 4, although better mixed and combined than the other, are not a complete combination. If they are melted again, their mixture becomes more intimate, their color more uniform, their fracture cleaner and not so easily effected, and their resistance greater.

A small proportion of lead with pure copper, as is the case with the alloys of copper and tin, copper and zinc, renders these metals more ductile, and better prepared to be rolled.

The proportion of copper 50, lead 50, may give an economical alloy, melting at a low temperature, compared with that required to melt copper, and which may be laminated, and found serviceable for those uses where hardness is not the main desideratum.

In order to obtain the alloys of copper and lead by the direct process, it is proper to heat the copper at the highest temperature which will not produce oxidation, then to add the lead already melted and raise the temperature during the stirring in the furnace, and, at last, to stir again just before running into the mould.

Generally, for these alloys made on a large scale, into which it is desirable to introduce lead, it will be proper to prepare in advance the alloy of equal parts of lead and copper, which appears to be the best suited for mixtures, then to employ this alloy to be remelted, whether with copper or with lead, according to the desired proportions.

8TH. ALLOYS OF COPPER, TIN, AND ZINC.

No. 1. *Copper* 80, *tin* 15, *zinc* 5.—Texture, a light violet.—Polish, a pale yellowish-pink, with the lustre of pure copper.—Fracture like that of red bronzes, half granular, half lamellar, and quite difficult to produce. —Resisting.—Malleable.—The surface of the button is like that of the bronzes of copper and tin.

No. 2. *Copper* 90, *tin* 8, *zinc* 2.—Texture, a very light greenish-yellow.—Polish, a light yellow with lustre.—Fracture, dry, a white ground, very slightly granular, and without lustre.—Very easily broken; hard to file; and very unyielding under the punch.—Its appearance is more like that of alloys of copper and zinc, with a large proportion of zinc, than that of alloys of copper and tin.—The surface of the button is covered with a wrinkled and light brown pellicle.—This alloy appears to be more sonorous than any of the others.

No. 3. *Copper* 75, *tin* 5, *zinc* 20.—Texture, a light greenish-yellow, sliding to green more than No. 2.—Polish, a light greenish-yellow, more easily tarnished than the preceding.—Fracture without lustre, striated towards the centre, and colored of a very light yellow tint, tending to white near the edges.—More resisting than the preceding, not so hard under the file and the punch, but quite dry and easily broken.—The surface of the button is smooth, brownish-yellow, and slightly concave in the middle.

No. 4. *Copper* 92, *tin* 2, *zinc* 6.—Texture, a light violet, although the tint is darker than No. 1.—Polish, a pale red reminding us of that of pure copper.—Fracture, granular and orange-yellow.—Tough, and difficult to break.—Tenacious.—Malleable.—Yielding to the punch.—Clogs the file a little.—The surface of the button is smooth, raised at the edges, having a brown tint tending to black, and presenting in the middle a scoriated appearance like that of pure copper buttons.

No. 5. *Copper* 80, *tin* 5, *zinc* 15.—Texture, a dirty yellow, tending to green less than No. 3, and more than No. 2.—Striated fracture, finer than that of Nos. 2 and 3, yellow in the centre and white at the edges.—More resisting than Nos. 2 and 3, more easily filed and more yielding to the punch.—It bends before breaking.—The button is smooth, and covered with a brownish-yellow pellicle.

No. 6. *Copper* 34, *tin* 33, *zinc* 33.—Texture, a dirty gray.—Polish, a dead white, without much lustre.—Smooth fracture, with a few laminæ possessing a certain brightness.—Very easily broken, and may be pulverized under the hammer.—Dry to the file, the filings being very fine, without clogging the tool.—Will not receive the mark of a punch without breaking.—The button is covered with a very wrinkled skin, of a dirty gray, with a few specks of oxide of zinc.

No. 7. *Copper* 20, *tin* 60, *zinc* 20.—Texture, a gray color, not so dark as that of No. 6.—Polish, whiter and with more lustre than No. 6.—Fracture, more granular and more jagged at the same time, with a dull white color, excepting a few specks slightly brilliant.—Softer, and adheres more to the file.—Yields more to the punch.—The button is more even than the preceding, and is covered with a skin of a dirty gray, tending to white, on account of the presence of oxide of zinc.

No. 8. *Copper* 20, *tin* 20, *zinc* 60.—Texture, like No. 6.—Polish, a dead white as dull as that of No. 6.—The same brittleness and resistance to the punch.—The fracture shows brighter spots, of a bluish-gray white, more perceptible than with No. 6.—The button has the same appearance.—Somewhat dryer under the file, the filings being as fine and brittle.

Nos. 9, 10 and 11. *Copper* 20, *tin* 40, *zinc* 40.—*Copper* 10, *tin* 45, *zinc* 45.—*Copper* 2, *tin* 49, *zinc* 49.—These numbers give samples which possess a great analogy with those of Nos. 6, 7, 8, and 12, as regards the texture and exterior qualities.—They are brittle white metals, without probable uses in the arts.—No. 11, however, bears some resemblance to number 13 of the alloys of copper and tin, and copper and zinc, in this particular, that, the proportion of copper being sensibly lessened, the peculiar qualities of the other metals predominate, and occasion a more serviceable combination than those where the proportion of copper is more considerable, as in the Nos. 9 and 10, for instance.

No. 12. *Copper* 50, *tin* 25, *zinc* 25.—Texture a dirty gray like Nos. 6 and 8.—Polish, a pallid white with very little lustre, which immediately disappears.—Square and smooth fracture, very bright, without any grains, facets, or striæ.—Very brittle, and easily broken to a fine and dry powder under the hammer.—Does not bear the action of a punch without breaking.—This alloy is more brittle than glass; the preceding six numbers are also very brittle, but not so much so as this latter.—They break under the hammer. No. 6 and No. 8, especially, when crushed, do not fly, but form a kind of cake full of rents.—On the contrary, No. 12 becomes reduced to a dry powder, without any appearance of cohesion.

General Observations.—The same as with the alloys of the two preceding series, the combinations of copper, tin, and zinc give products the more tough, malleable, colored, easily filed and turned, as the proportion of copper is greater.

The alloys become white, dry, hard, and brittle, when the proportion of copper is below two-thirds of the whole mixture. The compounds, where copper enters as one-half, are extremely hard and brittle. A remarkable fact is, that the alloy of half and half copper and tin is dry, brittle, and difficult to file; whereas the alloy of half and half copper and zinc keeps a certain coloration, may be filed, and, although brittle, possesses a certain amount of resistance. On the other hand, the alloy of copper 50, tin 25, and zinc 25, where copper also enters as one-half of the compound, is sensibly worse than the preceding two. This alloy is exceedingly brittle, is crushed under the smallest pressure, and seems to have retained none of the characteristic properties belonging to the component metals.

The alloy of equal parts of copper and lead, of all the various alloys which we have examined, is that which, with copper as one-half of the compound, appears to us the more serviceable; and that notwithstanding the difficulty shown by lead to become alloyed with copper or with zinc.

If we pass the alloys of copper and lead, not employed in the arts up to this day, and which to our mind might be serviceable for rolling, the alloys of copper and zinc are those to be preferred, because they allow of a smaller proportion of copper in the compound, without greatly impairing its qualities.

The alloys of copper and zinc admit of a proportion of from 35 to 40 per cent. of zinc, without entirely losing their tenacity, color, and the property of being easily filed; whereas the alloys of copper and tin, with an equal proportion of tin, are white, very brittle, and cannot rank among the metals which are to be filed and chiselled.

If equal parts of tin and zinc are added to copper, so as to form one-third of the alloy, this will be more resisting and stronger, and will be easily chiselled, although the chips are brittle and fly readily. This composition is the extreme limit of ternary bronzes useful in the arts. These more favorable results, in the same proportional limits as those given by the alloys of copper and zinc, and copper and tin, seem to be in contradistinction to observed facts, when the alloys, instead of being composed of two-thirds of copper, contain one-half only. Here is another proof of the curious transformations of metals when in the state of alloys.

Therefore, the most serviceable series of the ternary alloys of copper, tin, and zinc are those where the proportion of copper is not less than two-thirds of the compound. They comprise the most advantageous alloys for the casting of statuary bronzes.

It is well known that statuary bronzes require special qualities. Above all, they must possess sufficient fluidity to completely fill the moulds, and at the same time they must be adapted to the work of the file and the chisel. The combinations which appear to us as fulfilling these requisites, and which at the same time present various tints as required by the arts, may be classified thus:—

	Copper.	Zinc.	Tin.	
No. 13	84	11	5	Polish, yellow-red.
" 14	83	12	5	Polish, yellow-red.
" 15	81	15	4	Polish, orange-yellow.
" 16	78	18	4	Polish, orange-yellow.
" 17	73	23	4	The same, but lighter.
" 18	70	27	3	Polish, light yellow.
" 19	65	32	3	Polish, light yellow.

No. 13 is the limit of the reddish-yellow bronze,* and No. 19 of the light yellow one.

Nos. 16, 17, 18, and 19 are evidently harder, and more difficult to be worked than the preceding three alloys; but they are less expensive, because they contain more zinc, and their specific gravities are sensibly lower.

When we consider the beauty and durability of the work, the three alloys Nos. 13, 14, and 15 are evidently to be preferred. They also take better the color of old bronze (*patine*).

Several of these alloys, besides being adapted to statuary, also present excellent qualities for pieces of machinery and for antifriction metals. Nos. 1, 2, 4,

* These alloys are rather brasses, if a bronze be an alloy of copper and tin, alone, or with other metals, but where the proportion of tin predominates over that of the other metals—copper excepted.
Trans.

13, and 14 are the best in this respect. A combination which, by its amount of copper, is similar to No. 12 of the alloys of copper and tin, and copper and zinc, appears to give very good results. Known under the name of "Feuton's alloy," its composition, which we shall again examine further on, is copper 5.50, tin 1.450, and zinc 80. By its hardness, color, and tenacity, it ranks with the alloy No. 12 of copper and zinc.*

Another ternary alloy, which resists ordinary friction well, does not become heated, and saves a great deal of lubricating material, is composed of copper 57, tin 28, zinc 15. It is of a slightly yellowish white, very hard, not malleable, and may be filed sufficiently well. Like the preceding, it is much cheaper than the bronzes of copper and tin only; and that is its greatest advantage.

In general, the series of the alloys which we have considered, nearly all give white antifriction metals, and are very economical. But it remains to be proven whether they will resist traction, torsion, compression, etc., as well as the bronzes with a preponderating amount of copper. This we doubt, and await thorough experiments to decide. But it is certain that, as regards beauty and good keeping in machinery, the true bronzes are much to be preferred.

9TH. ALLOYS OF COPPER, TIN, ZINC, AND LEAD.

No. 1. *Copper* 78, *tin* 2, *zinc* 18, *lead* 2.—Texture, a gray tending to yellow.—Polish, a light yellow, tending slightly to a red.—Fracture, jagged and without lustre. —Breaks with difficulty.—Hard to file.—Resisting under the hammer.—Possesses malleability and tena-

* We think that the true name is *Fenton*, instead of *Feuton*. Under that name are to be found in the trade several antifriction metals, without any copper in them, and none the better for that, as regards friction and durability; but they are more easily prepared, melted, and cast into or upon pieces of machinery.—*Trans.*

city.—Ductile.—The surface of the button is scoriated, and of a dirty gray color.

No. 2. *Copper* 75, *tin* 2.50, *zinc* 20, *lead* 2.50.—Texture, a gray with a few yellow and violet tints, and covered with a white oxide.—Polish, a gold-yellow, tending to green.—Jagged fracture of a gold-yellow, somewhat pallid.—More easily broken than the preceding.—More easily filed and polished.—A very fine lustre, when polished.—Presents a certain tenacity, malleability, and ductility.—The surface of the button is wrinkled, and of a brownish-yellow color.

No. 3. *Copper* 70, *tin* 10, *zinc* 10, *lead* 10.—Texture a dirty gray.—Polish, a pale yellow, without much lustre.—Fracture, gray, somewhat granular, but dry and easily tarnished.—Brittle, and harder than No. 1. Less resisting under the hammer than Nos. 1 and 2.—Very slight malleability.—Appears exceedingly well adapted for resisting friction, and for journal boxes. The surface of the button is covered with a very wrinkled skin, of a light brown color.

No. 4. *Copper* 25, *tin*, 25, *zinc* 25, *lead* 25.—Texture, a somewhat dull greenish-blue.—Polish, a silver-white without much lustre.—Dry fracture, with a certain brilliancy, and with a ground slightly granular.—Breaks very easily.—It is filed without difficulty, but clogs the tool a little.—Bears well the mark of the punch.—The surface of the button is a dull grayish-white, and covered with a large quantity of oxide.

No. 5. *Copper* 22, *tin* 26, *zinc* 26, *lead* 26.—Its texture, polish, and fracture present the same characteristics as the preceding alloy.—Breaks more easily, although more yielding under the punch, and clogging the file more.—The surface of the button is like that of No. 4.—The specific gravity is greater than No. 4.

No. 6. *Copper* 74, *tin* 1, *zinc* 10, *lead* 15.—Texture, a gold reddish-yellow.—Polish, a yellow sliding to orange-red, without much lustre.—The grains of the fracture are fine and regular, of a gold-yellow color.—Resists fracture.—Yields well to the punch.—Malleable and very tenacious.—Easily filed, without being too hard, or clogging the file too much—Presents all the characteristics of a good bronze.—The surface of the button is a dull brown-red, like that of all the alloys where the proportion of copper largely predominates.

No. 7. *Copper* 74, *tin* 10, *zinc* 1, *lead* 15.—Texture, gray tending to pale yellow.—Polish, a pale reddish-yellow, without much lustre.—Fracture, finely granular and of a light pink-gray, like that of a bronze made of copper 88 and tin 12.—More resisting than the preceding under the hammer; harder and dryer to the file.—Better as to resistance to friction, but not so fine a color.—Less malleable than No. 6.—The surface of the button is granular, and scoriated like the buttons of copper and tin bronzes.

General Observations.—There is little difference between No. 1 and No. 2; the latter, however, has a finer color, and is better adapted to gilding and chasing. No. 1 is harder, more resisting, tougher, and better for friction surfaces than No. 2.

No. 3, without possessing the qualities of resistance, malleability, and mildness of Nos. 1 and 2, may give a good and economical bronze for certain pieces of machinery; but it will not suit for statuary work.

Nos. 4 and 5 offer this singular property, of being very brittle and soft at the same time; they are to be ranked among the white alloys, without sonorousness, and nearly useless for the arts.

Nos. 6 and 7, on the other hand, may be applied very advantageously. No. 6 is redder than No. 7, and also more malleable and not so dry under the file. It seemed to us not so resisting under the punch, which may be accounted for by the volatilization of part of the zinc.

No. 7 is worth, in appearance, the ordinary bronze for machines (copper 88, tin 12); and, on account of the lead, it is more tenacious, less brittle, and more economical. Experiments, made on a large scale, appear to confirm all these advantages.

It will be easily understood that we could not, without inconvenience, multiply the examples of these quaternary alloys. There are so many combinations possible between the four metals with which we have operated, that we have been obliged to confine ourselves to stating a few results only. Several intermediate trials, ranging within the limits of the alloys which we have indicated, went to confirm the fact, already pointed out in the alloys of tin, zinc, and lead, that the lead sensibly improves the nature of the alloys into which it enters in a small proportion. Thus, when the alloys of copper and zinc, or copper and tin, become dry and brittle, they may be modified, and acquire body by the presence of lead. The same alloys, holding a large percentage of copper, and indicated as being malleable, ductile, tenacious, etc., will, with the aid of lead, maintain these qualities through the rollers and the draw-plate. It is thus that a portion of lead, as small as 0.50 per cent., gives the best alloys for drawing out under the hammer, for sheets and fine wires; these alloys being composed of copper 67, zinc 32, lead 0.50, and tin 0.50.

In the quaternary compounds lead combines better than in its binary compounds with copper, or even than in its ternary combinations with copper and tin, or zinc. This is a remarkable fact to state. Besides, the presence of lead does not appear to essentially modify the external nature of the alloys of copper, tin, and zinc; and if it does not always impart important qualities for the use, the appearance is at least improved. At all events, the addition of lead is very economical.

These last observations are especially applicable to those combinations demanded by industrial constructions. It is certain that an addition of lead to the statuary bronzes which we have mentioned, will improve the nature of the products.

The Romans composed the bronze for their statues of copper 99, tin 6, and lead 6.*

The brothers Keller, who were so celebrated as bronze-founders, made their alloys with copper 91.40, zinc 5.53, tin 1.70, and lead 1.47.

The composition for the Vendôme column was copper 89.16, tin 10.24, zinc 0.498, and lead 0.102.

At last Mr. Darcet, who has made numerous trials, recommends the following two alloys as being the best adapted to gilding, chasing, and turning:—

Copper 82, zinc 18, tin 3, lead 1.50.
Copper 82, zinc 18, tin 1, lead 3.

All of these results prove that the quaternary alloys of copper, tin, zinc, and lead give the best bronzes for the founders of artistical castings. And this will be confirmed by any examination of Nos. 1, 2, 3, 6, and 7, besides many alloys which we do not mention. We shall add, however, that in these compounds a proportion of lead of over 3 per cent. takes somewhat from the fluidity of the alloy, prevents it from reaching the sharp angles of the moulds, and appears to prevent a good bronzing (*patine*) or gilding.

2. Alloys of Iron with Copper, Zinc, Tin, and Lead.

As we have already stated, the alloys of iron have not, up to the present time, neither by us nor by other persons, been studied with sufficient accuracy to present interesting facts for the arts, and, above all, to bring

* The ancients rarely employed zinc in their alloys.

out new results, susceptible of wide and practical application.

As a rule, iron may be alloyed with most metals; but its alloys, always difficult to effect, and in the majority of cases only with a small proportion of iron, have up to the present time resulted in very few applications to the arts.

It is evident that iron, introduced in small proportions into certain metals or certain alloys, will impart to them new and important qualities. However, the experiments thus far made have been hindered by difficulties in the preparation, which have removed all the interest felt for them, and sometimes rendered them entirely useless.

Besides the alloys of iron with the above-named metals, of which we shall indicate the principal known data, this metal has recently been experimented upon, in order to combine it with certain modern metals, such as tungsten, for instance, for the manufacture of fire-arms. But all of these attempts, which we consider more or less fruitless, and of which we shall speak further on, are not appropriate in this part of the work.

Alloys of Iron and Copper.—The alloys of iron and copper are difficult to produce, at least by the direct process. The copper remains in a pulverulent state within the iron, has a tendency to become precipitated to the bottom of the fluid mass, or in the moulds, and the combination is generally incomplete.

With certain precautions, and by operating gradually with small quantities of the metals, in order to make preparatory alloys, which serve afterwards to make the definitive alloy on a larger scale, it is possible to arrive at a union of iron with copper, that is rather a mechanical mixture than an alloy. The copper always shows its presence in the cast iron, and is easily seen in the grayish fracture, with grains without lustre, of cast iron mixed with copper.

No matter how small the quantity of copper mixed with cast iron, the latter is rendered dry, hard, and brittle. It is sufficient that a few particles of copper should become scattered in a bath of molten iron to render cold short the iron puddled from that cast iron.

This is the result of observations made by metallurgists and founders who have accidentally seen a small proportion of copper mixed with cast iron.

An iron holding any copper cannot be welded; it breaks under the hammer, and runs off at a temperature much below that necessary to burn an iron free from copper.

An alloy of copper 20 parts, and cast iron 1 part, gives a tough metal, hard, resisting, as ductile as copper, and presenting a fracture where the presence of cast iron can scarcely be ascertained.

An alloy of copper 10 and cast iron 1, becomes harder and dryer than the preceding. The metal is scoriated, full of holes, and seems to be wanting in cohesion. It may be forged when cold, and remains quite ductile, although we doubt whether it would bear the drawing process, which, however, we have not tried.

An alloy of copper 1 and cast iron 20, shows the presence of the copper in the whole mass. The cast iron, however, has become harder and more resisting.

This hardness and resistance, on account of the little homogeneousness of the alloy, do not appear as if they would be capable of utilization in the arts. Several authors have claimed that pig-iron, intended for castings, and holding 1 per cent. of copper, will become more fluid and tenacious, and will produce sharper castings. This result might be possible if the alloy were thoroughly made, with the copper uniformly divided throughout the mass. But when in the manufacture we throw aside the precautions possible in a laboratory experiment, one of these two things will happen: the copper is oxidized, and most of it becomes mixed with the scoriæ on top of the bath; or it becomes precipitated, and will be found on the surface of the castings in the shape of drops, spots, or scoriated deposits.

Alloys of Iron and Zinc.—The alloy of these two metals has been, up to the present time, so difficult to produce in a practical way, that it is of no advantage in the arts.

Although the specific gravities of the two metals are not very different, the great tendency of zinc to volatilize as soon as the temperature is raised a little above that of its point of fusion, prevents its union with iron, which requires for its fusion a high temperature.

It is true that in nature we find certain ores where calamine (carbonate of zinc) is united with iron; where tin and copper pyrites contain iron; and where there are also certain ores of iron combined with those of lead or zinc, &c.; but none of these primitive combinations appear to be capable of producing alloys, at least by the known processes.

When zinc is dipped into molten iron, it decrepitates, becomes divided, and is projected out of the bath in the shape of cadmiæ, without leaving a trace of its presence in the castings made after this attempt to alloy.

By means of peculiar precautions, we have been enabled to introduce zinc into molten cast iron, without, however, producing a regular alloy that could find a place in the arts.

Our process was to introduce a well-heated iron tube, down to a certain depth, into a bath of molten cast iron covered with a thick layer of charcoal-dust, and then to pour the melted zinc through that tube.

A part of the zinc was lost, but enough remained to form an alloy.

This alloy was hard, dry, and of a dull white color; it was also brittle when the proportions were approximatively zinc 50, cast iron 50. By increasing the quantity of zinc, the alloy became whiter, more like the texture of silver, and slightly more malleable. But, no matter what were the proportions of zinc or cast iron, the compound did not appear of any use in the arts.

The union of iron and zinc is possible by analogous processes to those employed in the manufacture of tinned iron. A well-scoured sheet of iron, plunged into a bath of molten zinc, becomes uniformly covered with a layer of the latter metal, and the adherence is sufficiently great.* However, we do not here arrive at results so good as can be had by the union of tin, or tin alloyed with zinc, for making tinned iron.

At the present time, it is by the processes of galvanization that we arrive at the best union of iron with zinc.

We shall indicate here, only as a memorandum, a few attempts made in the experimental laboratory at alloying zinc and iron.

These alloys have been experimented upon by reducing together the oxides of iron and of zinc, by cementing in charcoal-dust a mixture of oxide of iron and calamine; or by heating together in a well-closed crucible a mixture of cast-iron filings with granulated zinc.

All of these purely scientific processes gave no practical results; and, in the absence of new and more satisfactory trials, we are obliged to admit that cast iron is not at all improved by the addition of zinc, even in

* This adherence is even greater when the sheet-iron has been covered with lead, before being galvanized with zinc.

minute proportion; whereas, on the other hand, a zinc holding a small proportion of iron is more brittle, less ductile, and, in a word, inferior to zinc free from iron.

Certain iron ores, especially in Belgium and the North of France, contain a small proportion of zinc, which in a few cases may be collected from the blast-furnace. This zinc has no well-marked effect on the nature of the cast iron; although it is admitted, when the proportion of zinc is considerable, that the cast iron is dryer, more brittle, and more difficult to refine than that obtained from ores without zinc.

Alloys of Iron and Tin.—But for the great difficulty of operation resulting from the high point of fusion of iron, this metal might be alloyed in all proportions with tin. The specific gravities of the two metals are sufficiently alike to enable a good alloy to be produced.

These alloys, however, are brittle, and are the more difficult to melt as they contain more iron. With a high temperature, the alloy is easy, but there is a greater or less waste of tin.

A small proportion of iron in tin, gives to this metal a dull appearance, a greater hardness, and less malleability. On the other hand, a very small quantity of tin in iron renders it both cold and hot short, especially the latter. An iron holding a certain amount of tin cannot be forged, and flies to pieces under the hammer.

Cast iron which contains tin may present at its fracture as fine a grain as that of steel. It becomes black, and may acquire, like most of the hard metals, a fine polish not so easily tarnished as that of ordinary cast iron. Various attempts have been made in order to prevent the oxidation of cast iron by the addition of a small proportion of tin.

Our own studies have shown that, by doing so, the cost of cast iron will be increased by a greater difficulty to work it, due to its greater hardness, without imparting to it the necessary qualities for resisting oxidation successfully.

A proportion of 2 per cent. of iron in tin is sufficient to render the latter metal magnetic, hard, dry, and without lustre.

The same proportion of tin in cast iron, renders the metal dry and brittle, and the iron puddled from that pig-iron is hard and less malleable.

An alloy of iron 30, and tin 70, presents a dark gray fracture, a certain ductility, but nothing useful in the arts.

An alloy of iron 50, and tin 50, is white, brittle, and possesses a granular fracture.

An alloy of iron 70, and tin 30, is crystalline, with an iron-gray texture, and may be pulverized under the hammer.

An alloy of iron 90, and tin 10, is of a light gray shade. The grain, which is dry and without lustre, is filed with great difficulty. This alloy is very dry, and very brittle and hard. Its polish, obtained upon a stone, is of a grayish-white and fine lustre.

The practical uses for the combinations of iron and tin are the tinning of metals, which processes consist rather in a covering than in an alloy.

Tinned sheet-iron, which is often considered as an alloy of tin and iron, is nothing but iron covered with several layers of tin. The first layers may, possibly, form an alloy.

It is not our object to give the processes for tinning sheet-iron. We shall only mention that the main point is to produce a perfect adherence between the tin and the iron, and not a thorough combination, which would render the latter metal brittle. It is just on account of the penetration of tin, which we try to avoid, that there is no true alloy formed during the manufacture of tinned iron. We may, therefore, admit that tinned iron is made of a sheet of iron, a superficial rather than a complete alloy of tin and iron, and several layers of tin.

The sheet-iron for this manufacture must be of the first quality; and this quality should not be altered by the operations of pickling, scouring, and tinning.

With certain qualities of tin, some manufacturers add a small proportion of copper, in order to give more fluidity to the tin, which will then leave on the surface of the iron thin and regular layers.

The tinning of cast-iron vessels is even less of an alloy than the tinning of sheet-iron. Unless we use a very porous gray metal, there is no penetration by tin, and the tinning process in this case is but a covering with tin, the adherence of which to the cast iron is more or less complete.

For tinning copper, for instance, some employ an alloy of iron 10, and tin 60, made by fusing block-tin with iron scraps, and keeping the molten mass at a red heat for a certain length of time.

This alloy, which is very brittle when hot, possesses a certain malleability when cold. It is cut and filed with difficulty. Its fracture is gray, and finely granular.

Thénard has proposed, for the same purpose, an alloy holding less iron than the preceding, and composed of iron 10, and tin 80 parts. This alloy is grayish-white, fusible, denser and not so hard as the alloy of iron 10 and tin 60.

Alloys of Iron and Lead.—Equally so with zinc, we cannot produce, in a practical way, alloys of iron and tin which will be serviceable in the arts.

Lead, which is often difficult to alloy with other metals, unless employed in small proportions and with many precautions, has no affinity for iron.

A piece of lead thrown into a bath of molten iron, becomes oxidized, or is separated and found at the bottom of the bath after the cast iron has been run out. As soon as the lead is introduced into the molten cast iron, a certain agitation appears at the surface and even through the whole bath, and the cast iron seems more fluid. When thin or large pieces are to be cast, the founders who are aware of this phenomenon often throw a certain quantity of lead into the molten cast iron, in order to prevent it from congealing too soon against the sides of the casting-ladle.

The want of affinity of iron for lead, and conversely, is made use of for separating lead from other metals having a greater affinity for iron. On the other hand, lead may be employed for separating iron from other metals, such as silver, for instance. Thus, if lead is added in sufficient quantity to a fused alloy of cast iron and silver, it will combine with the silver, and the iron will swim at the surface of the bath.

All the authors who have occupied themselves with the question of alloys, agree upon the impossibility of alloying lead and iron.

In experiments made by ourselves at Angers, 1847–1848, we obtained a kind of saturation of iron by lead in certain mixtures thoroughly stirred, and rapidly cast, where the proportion of lead was not over 2 to 3 per cent. In all these experiments, whether because most of the lead was oxidized, and therefore could not be found in the trial bar, or because it was deposited in the shape of drops at the bottom of the moulds, it was ascertained by analysis that only traces of lead could be found. Which shows that lead had traversed the metal, without producing a true alloy.

The cast iron thus treated was harder, and its grains were flattened and without lustre. Its specific gravity was 7.2, which corresponds to the average of ordinary cast iron.

II.

ALLOYS OF THE METALS OF SECONDARY IMPORTANCE IN THE ARTS.

We shall successively examine the metals of the second series in the order of their alloys with the metals of the first series, and then between themselves.

Our observations shall be short. All these metals, up to the present time, have seldom been experimented upon, and that without method or perseverance. With the majority of these alloys, we find that the most conscientious workers entirely disagree. The facts which we indicate in this chapter sometimes result from our own observations; but we must confess that we have not had the time to make with these metals so conclusive and numerous experiments as with those of the preceding series. Therefore, we have been obliged to borrow occasionally from authors who, like ourselves, have examined these alloys more from traditional data than from well-verified experiments.

Alloys of Bismuth and Copper.—These alloys are easily effected, notwithstanding the difference in the points of fusion of the two metals. They are brittle, and of a pale red color, whatever the proportions employed. The specific gravity of the alloys is sensibly equal to the average of the two metals.

Alloys of Bismuth and Zinc.—These alloys are seldom made, and produce a metal more brittle, presenting a larger crystallization, with less adherence, than zinc or bismuth taken singly. On that account they are useless in the arts.

Alloys of Bismuth and Tin.—The combinations of bismuth and tin take place easily, and in all proportions. A very small quantity of bismuth imparts to tin more hardness, sonorousness, lustre, and fusibility. On that account, and for certain applications, a little bismuth is added to tin in order to increase its hardness. However, bismuth being easily oxidized, and often containing arsenic, the alloys of tin and bismuth would be dangerous for the manufacture of certain domestic implements, such as culinary vessels, pots, etc.

The alloys of bismuth and tin are more fusible than each of the metals taken separately.

An alloy of equal parts of the two metals is fusible, according to several authors who disagree, at a temperature varying from 100° to 150° Centigrade. These differences are evidently due to an incorrect measuring of the temperature, or to the temperature being taken after the proper time of fusion.

When tin is alloyed with as little as 5 per cent. of bismuth, its oxide acquires the peculiar yellowish-gray color of the bismuth oxide.

According to Rudberg, melted bismuth begins to solidify at 264°, and tin at 228°. For the alloys of the two metals the "constant point" is 143° C.

Alloys of Bismuth and Lead.—These two metals are immediately alloyed by simple fusion, with merely the ordinary precautions. The alloys are malleable and ductile as long as the proportion of bismuth does not exceed that of lead; they are also much more tenacious than lead.

The alloy of bismuth 2, and lead 3 parts, is about ten times harder than pure lead.

The compounds of bismuth and lead generally have a dark gray color, with a tint intermediate between the color of tin and that of lead. Their fracture is lamellar, and their specific gravity greater than the mean specific gravity of either metal taken singly.

An alloy of equal parts of bismuth and lead has a specific gravity equal to 10.71. It is white, lustrous, sensibly harder than lead, and more malleable. The ductility and malleability diminish with an increased proportion of bismuth, while they increase with the excess of lead in the alloy.

An alloy of bismuth 1 and lead 2 is very ductile, and may be laminated into thin sheets without cracks. Berthier says that its point of fusion is 166° C.

According to Rudberg, melted lead beginning to solidify at 325°, the "constant point" for the alloy of the two metals is 129° C.

Alloys of Bismuth and Iron.—The learned disagree as to the possibility of combining bismuth and iron. Up to the present day, the combinations indicated are rather doubtful.

At all events, the principal fact is, that the presence of bismuth in iron tends to render this metal brittle, and is not an improvement in its manufacture.

Alloys of Bismuth and Antimony.—These alloys are grayish, brittle, lamellar, like the alloys of bismuth and zinc, and present no real utility in the arts.

Alloys of Bismuth and Nickel.—As with the preceding combinations, we are not aware of any interesting application of the alloys of bismuth and nickel.

Alloys of Bismuth and Arsenic.—These alloys are more brittle and more fusible than bismuth. This metal, which is found in nature combined with arsenic, appears to have little affinity for it, when we make alloys. Nothing practical has been accomplished in the alloys of bismuth and arsenic. Arsenic is rapidly volatilized, and the very small proportion which is absorbed by bismuth is easily oxidized. Therefore the many difficulties attending the formation of the alloy, which itself presents little interest, have prevented further examinations.

GENERAL OBSERVATIONS.—It will be seen from the preceding data, that the alloys of bismuth are not at the present time important in the arts, excepting the fusible alloys made of bismuth and certain white metals, such as tin, lead, &c. The alloys of bismuth with tin, the latter predominating, seem to be the most interesting. The great fusibility of the alloys of bismuth and lead will have the effect of popularizing these alloys, and also those with tin, as soon as bismuth can be obtained in abundance and at a less cost.

To sum up, the action of bismuth in alloys is to increase their hardness, fusibility, and brittleness. But, although bismuth renders brittle the metals with which it combines, it does so a great deal less than arsenic or antimony, for instance.

Alloys of Antimony and Copper.—These two metals rapidly combine by fusion. Whatever are the proportions, and especially when antimony predominates, the alloys are brittle, of a violet color, and with a specific gravity above the average one of the two metals, considered singly.

The alloy by equal parts, which was named by the ancients *Regulus of Venus*, is of a grayish-violet color, which tends to a nearly pure violet when the proportion of copper increases within certain limits.

An alloy of antimony 1 and copper 3 seems to possess the violet shade to the utmost degree. It is dry, brittle, lamellar, more fusible than copper, and has a fine lustre when polished.

An alloy of antimony 1 and copper 6 is a reddish-yellow, having more of the copper than of the violet color. Its fracture is dryer, not so even, and more granular than the preceding one.

According to Mr. Hervé, author of a manual of alloys, antimony will whiten the copper with which it is alloyed more than is the case with an equal proportion of zinc.

Alloys of Antimony and Zinc.—The alloys of antimony and zinc are but little known. They are exceedingly brittle, and too easily oxidized by heat; their fracture is very lamellar and of a steel-gray color. They have presented but little interest to experimenters.

Alloys of Antimony and Tin.—The alloys of anti-

10

mony and tin are as white as tin, but harder and a great deal less ductile. They are the more brittle as the proportion of antimony is greater.

The specific gravity of these alloys is below that which would be calculated from the specific gravity of each metal, taken singly.

An alloy of tin 80 and antimony 20, although not so malleable as pure tin, is sufficiently so to be laminated and hammered when cold. It is by remaining near these proportions that the proper alloys of tin and antimony are made for the manufacture of tin pots and engraving plates.

Alloys of Antimony and Lead.—Antimony increases the hardness of lead, and renders it very brittle when the proportion of antimony is considerable. The alloy of lead 76 and antimony 24 appears to be the point of saturation of the two metals. More fusible than the average fusibility of the two component metals, ductile, and harder than lead, this alloy expands in cooling. To this property is due the employment of this alloy for the manufacture of type. But the above compound does not answer perfectly well, especially for small type. When too soft, it gets out of shape; when too hard, it cuts the paper; and it happens too often that the founder passes to one or the other extreme. When the alloy is melted in contact with the air, antimony is oxidized much before lead; and this accounts for the difficulty of obtaining an exact composition. It is a constant subject of study for type-founders, to arrive at a fusible and homogeneous metal, with much expansion, as resisting as possible, and at the same time soft enough to be repaired, and to bear the action of the press without being soon put out of shape.

The alloy of equal proportions is dry, porous, and brittle. These defects increase in the same ratio as the proportion of antimony. On the other hand, they disappear when the lead takes the place of antimony.

An alloy of antimony 1 and lead 4 is compact, much harder than lead, and remains malleable.

An alloy of antimony 1 and lead 8 possesses much tenacity, and a specific gravity greater than the proportional specific gravity of the two metals. It is more malleable than the preceding alloy, and retains a certain hardness. The hardness imparted by antimony, the increase of tenacity, and that of the specific gravity, are very perceptible up to the alloy of antimony 1 and lead 16.

Alloys of Antimony and Iron.—The two metals appear to have a mutual affinity. Their alloys, which are easily effected, are much more fusible than iron, and are white, hard, and brittle.

Their specific gravity is less than the average of the two metals. The alloy made of antimony 70 and iron 30 is quite fusible, white, and very hard. That made of antimony 30 and iron 70 is exceedingly hard, flies under the hammer, and produces sparks when filed.

Mr. Hervé has experimented with various alloys of antimony with cast iron. We cite the following:—

1 part of antimony.	100 parts of cast iron.
2 " "	100 " "
3 " "	100 " "

Antimony was added to the iron only when the latter was in fusion in the crucible.

The fracture of the samples of the first alloy was uneven, striated, and lamellar; the crystallization was confused, divergent, with a certain lustre, and grayish-white.

The fracture of the samples of the second alloy was, like the preceding, uneven, striated, and lamellar; the crystallization was confused, and of a grayish-white color, but duller.

The fracture of the samples of the third alloy presented the same characteristics as the preceding alloys, but the color was duller and darker. These samples like those of the second series, were very hard and brittle; square bars, with a side equal to 1.7 centimetre, were broken when falling on the ground, from a height equal to 1 metre.

Mr. Hervé has inferred, from these three experiments, that antimony is not entirely volatilized when thrown into fused cast iron, and that a portion remains in combination with the iron, on account of its affinity for the latter metal. On the other hand, antimony exerts a powerful influence on the crystallization of iron, during the cooling. One per cent. of antimony, at most, is sufficient to alter the fracture of cast iron, which then resembles that of zinc.

At all events, these alloys appear to be without application in the arts. They increase the brittleness of cast iron, whereas we always try to develop its tenacity. Cast iron may, it is true, thus acquire a little more lustre, when polished; but the advantage is so slight, that it does not warrant the increase of cost.

Alloys of Antimony and Nickel.—These alloys are brittle, of a lead color, and do not present any utility. They have not been studied.

Alloys of Antimony and Arsenic.—The two metals may be alloyed in every proportion. They combine with a production of light, and the resulting compound is, to a certain point, like the brittle and gray metallic mass found in the mineral kingdom, where native antimony is often found combined with arsenic.

The alloys of antimony and arsenic, which do not, however, present any interest, are very fusible, very hard and brittle, and present a fracture, with lamellar facets smaller and more characterized than those of pure antimony.

General Observations.—The useful alloys of antimony are those where this metal is combined with tin and lead. They are employed in the manufacture of types, engraving plates, and tin pots. The action of antimony is also of interest in certain ternary or quaternary alloys, where, like most of the metals used in such combinations, it tends to aid in the formation of a more complete and thorough alloy.

The ternary alloys of antimony, lead, and arsenic; antimony, lead, and bismuth; antimony, tin, and bismuth; antimony, copper, and lead; antimony, tin, and lead; antimony, lead, and zinc—have been, or are yet employed, some for the manufacture of types, others for that of ectypes or engraving plates. We, may then, say that antimony has been the indispensable, if not the predominating, base of all the alloys experimented upon for typographical or printing purposes.

Notwithstanding all these experiments, the known alloys are not perfect; and very likely a long period will elapse before we arrive at the best alloy for printing, that is to say, one fulfilling all the conditions of hardness, malleability, tenacity, expansion, and mildness, which have been found necessary by all those who have tried to improve the manufacture of types.

Amongst the quaternary alloys, where antimony has been found useful, we may cite the alloys of antimony, bismuth, copper, and tin; antimony, bismuth, tin, and lead; which have been, or are yet employed in the manufacture of the English metals called *pewter* and *queen's metal,* from which teapots and vases imitating silver have been made.

The following alloys of:—

Antimony, silver, copper, and zinc;
Antimony, tin, zinc, and steel;
Antimony, copper, iron, and lead;
Antimony, copper, tin, and zinc;
Antimony, copper, tin, and lead;

have been tried for the manufacture of metallic mirrors, buttons, and other products, when it was desirable to obtain a fine polish, a bright lustre, a certain hardness, and at the same time a sufficient mildness or malleability to allow of their being worked.

A remarkable property of antimony is to render brittle the metals with which it is united, even when it is in small proportion. This should be remembered, when we desire to employ this metal in experimenting on new alloys.

Alloys of Nickel and Copper.—The alloys of nickel and copper are easily effected by fusion. In the mineral kingdom, nickel is united with copper. In Piedmont, in the valley of the Sesia, are to be found large deposits of white magnetic pyrites, holding 5 per cent. of nickel and $1\frac{1}{2}$ per cent. of copper. In other countries, the nickel is to be found, under the name of white copper, amongst the slags of certain copper-works, where the nickel had been allowed to go to waste.

The alloy of 1 part of nickel and 2 of copper gives a grayish-white metal, slightly crystalline at the surface, tenacious, ductile, and sufficiently fusible.

Alloys of Nickel and Zinc.—Few experiments have been made on these alloys. According to certain chemists, Thomson for instance, nickel does not alloy with zinc by fusion. Others, on the contrary, assert that an alloy is possible, and they give as a proof, the use of it by the Chinese for the composition of their *pak-fong*, or white copper.

Berthier has tried to make an alloy of nickel and zinc; the resulting button had the composition of nickel 0.53 and zinc 0.47. It had a fine silver-white color, and could be hammered before it would crack and break. From this experiment, this skilful chemist infers that it might be possible to employ zinc for making, on a small scale, a melted nickel, compact and malleable. This is the most striking fact we have collected among the data, given by various authors, on the subject of alloys of nickel and zinc.

Alloys of Nickel and Tin.—We do not find any important experiments on these alloys. This remark applies equally to the *alloys of nickel and lead.* These alloys, however, are possible; and, if they do not appear immediately useful as binary combinations, they may become serviceable in the ternary and quaternary alloys, by introducing copper or zinc, or both, into the combinations of nickel and tin, or nickel and lead.

Alloys of Nickel and Iron.—Nickel easily unites with iron, and gives, according to certain authors, a soft and tenacious alloy. This fact is open to discussion. We may be allowed to suppose that nickel, like copper, has a tendency to render cast iron dry and brittle. Meteoric iron, and certain aerolites, contain from 3 to 10 per cent. of nickel.

This kind of iron, generally very soft when it is not combined with substances other than carbon, may acquire a very fine polish. It may be imitated by certain alloys of iron and nickel, which are less easily oxidized than iron, and remain ductile as long as the proportion of nickel is not over 10 per cent.

In England, MM. Faraday and Stodart have tried to reproduce meteoric iron with the following alloys. They melted in a crucible 97 parts of good iron and 3 parts of nickel: the alloy had the appearance of being as malleable and easily worked as pure iron. When polished, its color was quite white; the specific gravity was 7.804.

Another alloy of iron 90 and nickel 10 produced a metal having a yellow tint after having been polished, a specific gravity equal to 7.849, less oxidizable and malleable than iron, and more brittle than the preceding alloy.

An alloy of the same kind, tried by Berthier, by reducing in a brasqued crucible a mixture of oxides corresponding to 12 parts of iron and 1 part of nickel, gave a metal semi-ductile, very tenacious, with a fracture granular and slightly scaly, and presenting exactly the characteristics of meteoric iron.

MM. Faraday and Stodart have also succeeded in combining steel and nickel in the proportion of 10 parts of nickel with from 80 to 100 parts of steel.

But, in opposition to what has been previously related in regard to iron, they mention that steel, combined with nickel, is more easily oxidized than pure steel.

M. Dumas thinks that such alloys might be serviceable for the manufacture of telescopic mirrors, which quite contradicts the opinion that they are easily oxidized.

On the other hand, Karsten believes that the experiments of MM. Stodart and Faraday produced no true chemical combinations; and that, if a metal united with steel only by simple mixture, may increase its tenacity, the same effect may not take place with a thorough combination. Recent and not very conclusive experiments have been made in that direction, for combining cast iron and steel with wolfram (tungsten), in the hope that a resistance superior to that of the ordinary metals will be obtained.

Alloys of Nickel and Arsenic.—We mention these combinations rather as existing in nature than as a future source of useful alloys for the arts. According to Berzelius, nickel easily combines with arsenic, and holds it, even when submitted to a very high temperature. A small proportion of arsenic added to nickel does not impair the malleability or the magnetic property of the latter metal, but increases its fusibility. Alloys made under these conditions are very hard, and tinged with a light red shade. Their specific gravity, according to Thomson, is much below the average of the two metals.

GENERAL OBSERVATIONS.—The preceding indications show sufficiently well that nickel and its alloys have been submitted to sufficiently thorough investigations, in order to reveal unexpected facts. It is, however, sure that if nickel were produced in greater abundance, this metal would find new and useful applications. At the present time, the industrial uses of metallic nickel seem to be confined to certain alloys with copper and zinc, which, in Birmingham especially, are employed in the manufacture of white metal wares, imitating the color, lustre, and polish of silver. The proportions of these alloys remain in the neighborhood of copper 8, nickel 2 to 6, and zinc from 3_2 to 6. When the proportion of nickel is below 2 parts, the metal obtained is not better than a pale brass, and tarnishes rapidly in the air. When the proportion of nickel is 6 parts or more, the alloy possesses a fine polish with much lustre, but is difficult to produce, and subject to shrinkage, fracture, and other accidents during the casting.

Alloys of Arsenic and Copper.—It is difficult to combine directly copper and arsenic. This latter metal is not held with sufficient strength by the copper, whose high point of fusion volatilizes it before the combination can take place.

The alloy is obtained by melting copper and arsenic in a covered crucible, with a layer of salt or charcoal-dust, in order to prevent oxidation of the arsenic by the air.

The alloy of equal parts of copper and arsenic is white, brittle, and without malleability. It becomes slightly ductile and malleable only by considerably diminishing the proportion of arsenic. The contact of the air tarnishes it. By calcination, the greater part of the arsenic disappears by volatilization, and the remaining metal regains a certain malleability.

The alloys of copper and arsenic are generally known under the name of *white copper* or *tombac.*

The ordinary composition of these alloys is about

copper 62 and arsenic 37. They are of a brilliant gray color, very brittle, fusible at a red heat, and unaltered at the temperature of boiling water. By increasing the proportion of copper, the alloy becomes whitish, somewhat ductile, and is preferred for the manufacture of small articles of *white copper.*

Alloys of Arsenic and Zinc.—The alloys of these two metals are difficult of preparation. They are very brittle, and useless for the present wants of the arts.

Alloys of Arsenic and Tin.—These two metals easily combine by fusion, and in all proportions. The alloys are gray, lamellar, brittle, and less fusible than tin.

By its union with arsenic, tin becomes whiter, more brilliant, harder, and more sonorous; but it becomes very brittle if it contains but one per cent. of arsenic.

6 parts of arsenic with 100 of tin are sufficient to produce an alloy, crystallizing with large laminæ, like bismuth, and entirely deprived of ductility.

The alloys of arsenic and tin are of no actual utility in the arts. A compound of arsenic 1 part and tin 3 parts is employed in laboratories for the preparation of the arseniureted hydrogen gas. The arsenides of zinc may be used instead.

Alloys of Arsenic and Lead.—These combinations are not produced without difficulty, and not equally easy in all proportions. Beyond the proportions of arsenic 16 parts and lead 84 parts, which seem to be the highest degree for an intimate atomical combination, the arsenides, where the proportion of arsenic is greater, are easily decomposed by raising the temperature. The metal also becomes brittle, and presents a fracture like that of bismuth, but of a darker color.

The arsenides of lead are, therefore, the less ductile and the more brittle, as they contain more arsenic; their fracture is brilliant, lamellar, and of a grayish-white color. They are very fusible.

A white heat expels a notable portion of the arsenic,

and seems to leave an arsenide having the constant composition of arsenic 1 and lead 2, which will bear a very high temperature without losing weight.

The arsenide of lead is employed for facilitating the manufacture of shot lead, which is prepared, as we know, by letting fall from an elevated place drops of lead into water. An addition of two or three thousandths of arsenic to the lead helps its solidification, and gives to the shot a more spherical shape.

Alloys of Arsenic and Iron.—These alloys are possible in various proportions, but they have no direct utility in the arts. The alloy is more or less white, hard, brittle, and with a fracture resembling that of steel, with grains finer than those of iron. The most evident result of the combinations of arsenic with iron, and we may say with most metals, is that iron becomes harsh, brittle, and loses much of its malleability and ductility even with a very small amount of arsenic.

GENERAL OBSERVATIONS.—The alloys of arsenic, generally known under the name of arsenides, are rather "unions" than alloys of metals. Nevertheless, these "unions" possess a metallic lustre. The effect of the presence of arsenic is to increase the brittleness and fusibility of the metals with which it is united.

The arsenides having a certain importance in the arts are the alloys of arsenic and copper, known under the name of *white coppers.* Among the ternary and quaternary combinations, we may mention the following, which are more or less employed:—

Arsenic, antimony, and lead, for types.

Arsenic, bismuth, and copper, for buttons.

Arsenic, copper, and tin, tried for the manufacture of telescopic mirrors, and other optical instruments.

Arsenic, copper, tin, and zinc, also tried for telescopic mirrors.

Amalgams.—These are alloys of mercury and other metals; but we shall not dwell on these compounds, as they do not strictly belong to our subject, which comprises more especially those combinations obtained by fusion in the foundry.

The amalgams of *mercury and copper* are difficult of preparation, and present no practical interest.

Mercury and zinc give white compounds, very brittle, and remaining pasty when mercury predominates.

Mercury and tin combine in all proportions with the aid of heat, and will also combine at the ordinary temperature. The amalgam formed of mercury 10 parts and tin 1 is liquid, and resembles mercury, except that it does not run so well.

An amalgam of equal parts of mercury and tin is solid.

An amalgam of *mercury and lead*, half and half, is susceptible of crystallization. With the aid of heat, lead is very rapidly dissolved by mercury. At the ordinary temperature, the solution is effected by rubbing and trituration. Mercury may absorb half of its weight of lead, and yet remain liquid.

Mercury and iron do not directly combine. Mercury being without action upon iron, it is kept and transported in iron bottles or vessels. The amalgams of iron which are effected with the aid of potassium and zinc, or by any other indirect process, have no stability.*

Mercury and bismuth may form a kind of solution, by which mercury absorbs a great proportion of bismuth, without losing its fluidity; the drops, however, affect the pear shape. The amalgam of mercury 4 parts and bismuth 1 part is very fusible, and may be used for tinning, it being very adhesive to bodies with which it comes in contact.

The amalgams of antimony are granular, white, without consistence, and present no interest.

* The greater part of these data are borrowed from the interesting works of Mr. Berthier.—*Author.*

The same remarks apply to the *amalgams of nickel and arsenic.*

The amalgams which are most employed in the arts, and outside of those which belong to the laboratory, are those of tin for silvering mirrors, and the preparation of *mosaic gold;* and of tin or zinc for exciting electrical apparatus, &c. Mercury also enters into the composition of a few ternary and quaternary compounds, of which we may mention:—

A fraudulent amalgam of mercury 3 parts, lead 1 part, and bismuth 1 part, which is very fluid at the ordinary temperature, and is used for adulterating mercury. This alloy, which is fluid enough to pass through chamois leather like pure mercury, has its drops pear-shaped; which is a means of ascertaining the fraud.

The amalgam of Mr. Makenzie, which is solid at the ordinary temperature, and becomes liquid by simple friction, may be prepared as follows: melt 2 parts of bismuth and 4 parts of lead in separate crucibles; then throw the melted metals into two other crucibles, each containing 1 part of mercury. When cold, these alloys or amalgams are solid, but will melt when rubbed one against the other.

The amalgams of mercury, bismuth, tin, and lead, which are very fusible, are employed for metallic injections, and the silvering of the inside of glass globes and hollow mirrors, &c.

There is, in our opinion, no doubt that these metals of secondary importance in the arts, which we have just examined, will yet be called to take an important part in the practice of the industrial arts; and that several of their alloys will sooner or later emerge from the experimental state, in which, up to the present time, they have given only neutral results.

For this, it will be necessary, as with copper, zinc, tin, and lead, to take bismuth, antimony, arsenic, and nickel, and examine their combinations between each other, and afterwards those with the other metals. We do not here mention mercury, because this metal will require other kinds of experiments.

It will, therefore, be necessary to undertake a long series of comparative experiments, and not to abandon them until all the practical facts have been gathered. But such experiments are not easy, and cannot be conducted in a short time and without expense. They will consume much time and money, which are not at the disposal of everybody.

For us, these experiments would be very attractive. They were even a part of our programme when we undertook our first studies. But it is impossible to say that we will ever find the opportunity and the years necessary for their study.

III.

ALLOYS OF THE PRECIOUS METALS, BELONGING ESPECIALLY TO THE ARTS OF LUXURY.

The metals of which we will treat in this chapter are gold, silver, platinum, and aluminium. We shall consider them in regard to their combinations with the preceding metals, and between themselves.

We shall try to pass rapidly over the data which present no interest in the arts, leaving for the end of this book, where we sum up all the known alloys, the completion of what we have omitted or have not clearly explained.

Alloys of Gold and Copper.—Gold and copper have a great mutual affinity, and may be alloyed in all proportions.

The alloys are harder and more fusible than gold alone. Copper diminishes the ductility of gold, when it enters into the combination in a proportion over 10 to 12 per cent.

The great value of gold is the reason, in every case, why the proportion of copper in the alloy should not be very considerable. This remark applies equally to all the alloys of gold, and of the metals of this chapter, with all the common metals. We may hence observe, that the more precious a metal is, the less it should be mixed with common metals, in order not to be debased. Or, in other words, the higher the value of a metal, the greater should be the proportion of this metal in the alloy, unless in a very few cases, when a certain purpose is to be reached without reference to cost.

However, there are exceptions; as, for instance, when the costly metal being combined in a small proportion with a common metal, increases the value of the latter by imparting to it new and valuable properties. Such is the case with aluminium and copper. Aluminium, at the present time, is expensive, and therefore a precious metal. But when it is combined in a small proportion with copper, for the manufacture of the aluminium bronze, a new compound is produced which possesses many of the qualities of the precious metals, that is, lustre, brilliancy, solidity, and, above all, a great resistance to oxidation. We may, therefore, employ aluminium in small or large proportions, but we cannot do so in regard to gold and silver, which will be debased by a large admixture of other metals.

Gold, which is considered as the purest, most unalterable and perfect metal, must acquire a certain hardness, which alone it does not possess, for the manufacture of *coins, medals, jewelry, etc. It acquires that hardness* and solidity by being alloyed with copper. In such alloys, the respective proportions of gold and copper form what is called "the degree of fineness." Or, in other words, the fineness is the greater as the alloy contains more gold.

The standards or degrees of fineness are variable with different countries, and are regulated by law, especially those of the coin. We shall further examine this subject hereafter.

The specific gravity of the alloys of gold and copper is less than the average of the two metals.

An impure copper alters the malleability of gold, and may render it very brittle. A pure copper is therefore necessary for these alloys.

The maximum of hardness caused by the admixture of copper with gold appears to be when the alloy is made in the proportions of gold 7 parts and copper 1 part.

Alloys of Gold and Zinc.—The alloys of gold and zinc are greenish-yellow, brittle, and susceptible of receiving a brilliant polish. The zinc produces a sensible contraction in these alloys, and it so readily alters the qualities of gold, that when fumes of volatilized zinc reach melted gold, this latter metal becomes brittle.

An alloy of gold 11 parts and zinc 1 part resembles the pale yellow brass obtained with an excess of zinc. Its specific gravity is 19.937, and it does not tarnish.

Alloys of Gold and Tin.—The alloys of gold and tin are easily effected by fusion, and in all proportions. They are generally brittle; but may retain a certain ductility, when the proportion of tin is not over one-twelfth. The color of these alloys is pale and nearly white. Like the alloys of gold and zinc, the union of the two metals produces contraction.

Alloys of Gold and Lead.—These alloys may be produced in all proportions; they are exceedingly brittle, and without any utility in the arts. According to Berthier, one-half of one-thousandth of lead alloyed to gold is sufficient to render the latter metal entirely brittle, and without any ductility. All the alloys of gold and lead present the phenomenon of expansion, which is the greater when the proportion of lead diminishes, and its place is taken by copper, the proportion of gold remaining constant.

The maximum of expansion takes place when the lead is only 0.001 of the alloy.

An alloy of gold 11 parts and lead 1 part possesses the color of gold; but its fragility is such that it breaks like glass. Its fracture is finely granular, of a light brown, with a metallic lustre, and the appearance of broken chinaware.

The specific gravity of this alloy, which is harder and more fusible than gold, is 18.080, or a little less than the average specific gravity of the two alloyed metals.

Alloys of Gold and Iron.—Gold and iron easily combine in all proportions. Their mutual affinity is very great, and their alloy is decomposed with difficulty. Gold facilitates the fusion of iron, which is a proof of the tendency of these metals to become alloyed.

According to Karsten, iron does not change the tenacity of gold; and, on the other hand, gold does not seem to impair the qualities of iron, or to be an impediment to its manufacture.

An alloy of gold 1 part and iron 3 parts melts at a temperature below the point of fusion of iron.

An alloy of equal parts of gold and iron is of a grayish color, brittle, and somewhat magnetic. An alloy which contains $\frac{1}{12}$ of iron is pale yellow, and the color becomes grayish-yellow when the proportion of iron is increased to $\frac{1}{6}$. The alloy is grayish-white when the atomical proportions are 3 to 4 for iron and 1 for gold.

The alloy holding $\frac{1}{6}$ of iron is employed in jewelry, under the name of *gray gold*. The alloy where iron enters as $\frac{3}{5}$ or $\frac{4}{5}$ has been tried for making cutting instruments. This furnishes a metal susceptible of taking a hard temper.

Alloys of Gold and Bismuth.—The alloys are obtained by fusion, in all proportions; they have the appearance of brass, and are harsh and brittle. A trace of bismuth is sufficient for rendering gold brittle and without ductility. The action of bismuth on gold is the same as that of zinc. Vapors of bismuth, in contact with melted gold, are sufficient to impair the malleability of gold, and make it brittle.

The alloys all present the phenomenon of contraction; they are greenish-yellow, and their fracture is finely granular, with an earthy appearance. A compound made of equal parts of gold and bismuth has a specific gravity equal to 18.058, and suffers a loss in volume equal to 1.2 per cent., which shows a considerable contraction.

Alloys of Gold and Antimony.—Antimony possesses a remarkable affinity for gold, and dissolves it rapidly. The slightest fume of antimony is sufficient to alter the malleability of gold, and cause its brittleness. The alloys of gold and antimony are of a pale yellow color, and their fracture is finely granular, resembling that of chinaware.

The facility with which antimony unites with gold, attracted, from the earliest epochs of science, the attention of the alchemists, who pretended that gold increased in weight, when, after having been combined with antimony, the latter metal was separated. From that false idea, due to an imperfect separation, it had been supposed that antimony exerted a certain influence on the production of gold, and therefore favored the "transmutation." Thence the name of *Regulus* (little king) given to antimony, as characterizing the tendency of gold to assimilate this metal.

Alloys of Gold and Nickel.—To the best of our knowledge, these alloys have not been experimented upon. Some useful results might possibly occur by trying to substitute nickel for copper in certain gold alloys.

Alloys of Gold and Arsenic.—Gold easily combines with arsenic; the products are white or grayish-white, and very brittle. One-thousandth of arsenic is sufficient to take off all the malleability of gold, although its color is not changed. The arsenide of gold, prepared by exposing melted gold to the fumes of arsenic, is white and very brittle. Once united with gold, arsenic cannot be removed, except by a very high heat.

Amalgam of Gold.—Mercury has a very powerful action on gold; it dissolves it in large proportions, without losing its fluidity. The point of saturation appears to be 2 parts of gold for 1 part of mercury. The gold amalgam may be produced at a very low temperature, by the fumes of mercury. A piece of gold, rubbed with mercury, is immediately penetrated by it, and becomes exceedingly brittle.

The compound of gold and mercury is white, pasty, and crystallizes when cooled slowly. The amalgam, saturated with gold, is yellowish-white, remains soft, and may be kneaded between the fingers.

The great affinity of gold and mercury is the base of all the processes for gilding metals, especially copper. For gilding copper, bronze, and brass, we employ amalgams formed of 8 to 9 parts of mercury to 1 of gold. The uses of these amalgams have been greatly lessened since the adoption of the galvanoplastic methods. The description of the old process is to be found in a special treatise by d'Arcet; and we refer those of our readers who may be interested in the gilding process to that and other treatises on the subject.

Alloys of Gold and Silver.—Gold and silver may be easily mixed together, but do not appear to form true combinations. These alloys, more fusible than gold, do not seem to unite intimately, except in small proportions, and that without evident utility. Made within these conditions, the compounds are generally greenish-white, more ductile, harder, more sonorous and elastic than gold or silver, considered singly. One-twentieth of silver is sufficient to modify the color of gold.

Silver, like copper, increases the firmness and toughness of gold, and on that account it is employed at various degrees of fineness for jewelry work. These alloys are known by jewellers under the names of *yellow gold, green gold,* and *pale gold,* according to the proportion of silver. Green gold contains about 30 per cent. of silver; and pale or white gold, as much as 66 per cent.

Gilt silver is silver gilt with gold amalgams, and by processes of manufacture similar to those employed for gilding copper.

At all events, whatever is the temperature, the alloys of gold and silver are not susceptible of oxidation, whether by contact with the atmospheric air or with pure oxygen.

Alloys of Gold and Platinum.—The two metals may be alloyed in all proportions; but, on account of the infusibility of platinum, the alloy takes place only at a very high heat. All of these alloys are ductile and very elastic.

The combinations of gold and platinum have been studied by many chemists, who do not entirely agree, even on the appearance of the alloys. Some pretend that a very small proportion of platinum is sufficient to modify the yellow color of gold, and that an alloy made of 4 to 6 parts of gold to 1 of platinum possesses nearly the color of pure platinum. Others, on the contrary, claim that as long as the proportion of platinum is not over one-seventeenth of the alloy, the color of gold is not sensibly altered.

It would be interesting to ascertain the limit of modification in the color of gold, as, for instance, in the case of platinum fraudulently alloyed with gold. This fraud, which at a certain epoch was to be feared, does not appear to be extensively practised, and may be detected by the powerful means which modern chemistry possesses for determining and separating the most intimate compounds.

GENERAL OBSERVATIONS. — From the preceding data, we observe that gold is one of the metals which most readily enters into combination with other metals. But this property is without importance when we consider the inutility of the majority of the compounds, and the necessity of not debasing the value or impairing the qualities of gold. Moreover, it is certain that, excepting its alloys with copper, silver, iron, and platinum, the latter two being without actual utility, gold loses part of its ductility, resistance, and cohesion when it is combined with other metals, such as zinc, tin, lead, etc. Therefore it is entirely useless to experiment on those alloys where gold loses not only part of its money value, but also these valuable properties which participated in making it a precious metal. A similar reasoning will equally apply to the ternary or quaternary alloys into which gold may enter as a component part; they are not rational, and we shall not examine them.

Alloys of Silver and Copper.—Silver and copper are easily alloyed in all proportions. The combination takes place with expansion, and its specific gravity is less than that calculated from the proportions of the component metals. The majority of these alloys are as ductile as pure silver, and possess a great deal more hardness, elasticity, and sonorousness. The presence of copper does not modify the color of silver, so long as the proportion of copper is not above 35 to 40 per cent. A greater proportion of copper imparts to the alloy a yellowish tint, similar to that of brass; and if the combination contains from 65 to 70 per cent. of copper, its color is reddish, approaching the tint of pure copper.

The peculiar qualities of the alloys of silver and copper cause them to be preferred, in certain cases, and within certain limits, to pure silver, which is wanting in hardness.

An alloy made of 9 parts of silver and 1 of copper is white, tough, more fusible than silver, but not quite so ductile. These proportions are adopted for the French silver coinage. The maximum of hardness of the alloys of silver and copper appears to be when the proportion of copper is one-fifth.

These alloys of silver and copper, although easily effected by the ordinary processes of fusion, are nevertheless subject to the defect of separation or "liquation," which necessitates certain precautions when running the metal into the moulds. When such an alloy is run into a cold ingot-mould, the centre of the ingot is at a lower degree of fineness than the portions nearer the mould; and we observe, also, even in the monetary alloys, that all the portions are not of the same degree of fineness.

Alloys of Silver and Zinc.—Silver and zinc easily combine, but their products present no interest. They have a bluish tint, and a finely granular fracture. They are of no practical use.

Alloys of Silver and Tin.—The alloys of silver and tin present the phenomenon of contraction. They are harsh, very hard, and brittle. A small proportion of tin is sufficient to destroy the ductility of silver, and make it brittle.

Alloys of Silver and Lead.—Silver and lead unite in all proportions. A very small proportion of lead is sufficient to sensibly diminish the ductility of silver. The alloys are more fusible, and of a greater specific gravity, than the average of the two component metals.

An alloy of equal parts of silver and lead possesses

a great deal more of the properties of the latter metal than of the former. It is soft, and quite ductile and malleable.

An alloy of 1 part of silver and 7 parts of lead is grayish-white, less ductile than lead, and much less than silver. This alloy is less fusible than pure lead.

The alloys of silver and lead are easily and entirely decomposed by the process of cupellation. Lead plays an important part in the metallurgy of silver, which it is not our intention to consider here, our object being only the alloys proper.

Alloys of Silver and Iron.—We cannot say whether or not these alloys, which present no interest, have been the subject of any experiments.*

Alloys of Silver and Bismuth.—These alloys are very fusible, harsh, brittle, and lamellar. Their color is white, similar to that of bismuth. They possess a certain malleability, but are without interest. Bismuth is considered by a few chemists as being preferable to lead for refining silver; but its cost is too high for this purpose.

Alloys of Silver and Antimony.—These combine in all proportions, and the alloys present a whitish color tending to gray when they are overcharged with antimony. They are always brittle. Certain combinations of silver and antimony are found in the natural state, which possess a gray lamellar fracture, and a great brittleness and fusibility. They are easily decomposed by cupellation or fusion with nitre.

Alloys of Silver and Nickel.—There are no data on these compounds.

Alloys of Silver and Arsenic.—These alloys may be

* No true alloys of silver and iron have been made, only more or less intimate mixtures, where silver appears in the shape of drops, or filaments. The experiments of Messrs. Stodart and Faraday, made with steel, rather than with iron, show that the proportion of $\frac{1}{500}$ of silver corresponds to the best mixture.—*Trans.*

formed directly by fusion; and the silver will retain a certain proportion of arsenic even when the temperature is very high. The compound made of 86 parts of silver to 14 parts of arsenic is of a dead grayish-white color, brittle, and acquires a metallic lustre by friction. It is very fusible.

Amalgams of Silver.—Mercury acts upon silver with nearly as much power as upon gold. Therefore, silver amalgams are employed for silvering, in the same manner as gold amalgams for gilding.

The amalgam of silver is white, very fusible, and remains soft; its specific gravity is above the average of the two metals. It crystallizes easily, is not altered by contact with the air, and is dissolved only in a large proportion of mercury. It is decomposed by heat. This amalgam may be produced by throwing granules or scraps of silver into 12 to 15 parts of mercury, heated to 200° C. By pressing the product through a chamois-skin, the free mercury runs out, while the soft amalgam is retained, and used in that state for silvering.

We find in the mineral kingdom a crystallized silver amalgam, which is soft and possesses a very bright grayish-white lustre. According to Berthier, its specific gravity is 13.755.

Alloys of Silver and Platinum.—Platinum forms with silver a white alloy, which is harder and tougher than silver, and is less fusible and ductile as the proportion of platinum is greater.

This alloy is difficult to produce, on account of the separation of the platinum, due to the superior specific gravity of the latter metal. What might be its uses, we do not see, unless it is employed for the separation of gold from platinum, the alloy of platinum with a great excess of silver being soluble in nitric acid, while gold remains unaffected. This process may be useful for certain kinds of native gold, holding platinum.

GENERAL OBSERVATIONS.—The alloys of silver present a real interest only when they are made with gold or copper.

With the other metals up to the present time, and with very few exceptions, they are of no use in the arts. The alloys of silver and gold, and silver and copper, are those employed for articles of luxury, and for coin. The alloys of silver, gold, and copper are used for the same purposes. These ternary compounds are much used in England for coins and by goldsmiths. An alloy of silver, copper, and tin is made into a solder for plated ware and false jewelry.

The alloys of silver, copper, and platinum are also employed, but on a very small scale, for certain articles of jewelry and watchmaking. The quaternary compound of silver, copper, gold, and platinum produces an alloy having the appearance of the article known under the name of *doré* (gilt).

The alloy of silver, copper, tin, and gold is easily effected, and gives a tough and lasting metal. This alloy is found in certain coins and medals of antiquity.

The alloy of silver, arsenic, copper, and tin has been tried as speculum metal for telescopic mirrors. Equally with many other alloys, tried for the same purpose, this compound has not fulfilled expectations.

Alloys of Platinum and Copper.—These alloys are obtained by fusion, in all proportions. Like all the compounds into which platinum enters, they require a high temperature for their fusion.

The products vary with the proportion of platinum. With equal parts, the alloy is of a pale yellow, more brittle than malleable.

A compound of 1 part of platinum to 4 parts of copper is hard, although ductile, of a yellow pink color, and susceptible of a fine polish.

A compound of platinum 3 parts and copper 2 parts is nearly white, very hard, brittle, and without ductility. When the proportion of platinum is more than one-half, the alloy is sensibly hardened, and the color of copper rapidly disappears.

The alloys of platinum and copper, even with a small proportion of platinum, are much less oxidable than the alloys of copper with zinc or tin, for instance.

Alloys of Platinum and Zinc.—We know nothing relative to these alloys; moreover, they are scarcely possible by the ordinary processes of fusion, on account of the great tendency to volatilization in zinc, and the high point of fusion of platinum.

Alloys of Platinum and Tin.—These alloys take place in all proportions, but with the oxidation of a considerable portion of the tin employed, the alloy being formed at a white heat.

They are more or less brittle, or fusible, according to the proportions of platinum. A small percentage of the latter metal is sufficient to impair, and even destroy, the malleability of tin. An alloy of equal parts of platinum and tin is brittle, of a dark gray color, and with a coarse granular fracture. It tarnishes rapidly after being polished.

If the proportion of platinum is not more than one-tenth of the alloy, the latter becomes much more ductile, white, and lustrous, and its polish is much less easily tarnished.

Alloys of Platinum and Lead.—We possess no data on these alloys, which do not appear to have been experimented upon.

Alloys of Platinum and Iron.—By the ordinary processes of fusion, platinum appears to combine in all proportions, if not with wrought iron, at least with its carburized compounds, pig-metal and steel.

Berthier has tried alloys made of 1 part of platinum with from 4 to 10 parts of iron. The fusion was complete in brasqued crucibles. The fracture of the alloy was gray and granular, and it was possible to flatten the metal with a hammer before breaking it. The alloy was also easily filed, and presented a fine polish, whose tint was more like that of platinum than of iron.

These alloys appear to have remained within the limits of the laboratory, without having ever been employed in practice. The same may be said of the alloys of steel and platinum, which, however, have been the subject of more serious and conclusive trials on an industrial scale.

MM. Stodart and Faraday have made quite positive experiments on the alloys of platinum and steel.

With equal parts, they have obtained a metal which takes a very fine polish, not susceptible of being tarnished, and with a specific gravity of 9.862. With 90 parts of platinum to 20 parts of steel they produced an equally homogeneous alloy, which did not tarnish, and had a specific gravity equal to 15.88. Both alloys were malleable.

It would seem that platinum presents the advantage of removing from steel its tendency to become oxidized. This is the reason why the alloys of platinum and steel have been tried for certain weapons.

The best proportions for that fabrication appear to range between 2 and 3 of platinum for 100 of steel.

According to M. Dumas, an alloy of platinum 10 parts and steel 90 parts is very well adapted to the fabrication of mirrors. Its specific gravity is 8.1.

M. Bréant, inspector of the mint of Paris, had, about twenty years ago, caused to be tried several pieces of cutlery made of an alloy of ½ part of platinum to 100 parts of highly carburized steel.

The bulletins of the "Société d'Encouragement" have mentioned a few very remarkable samples. However, since that time, we do not believe that the usual applications of these products have followed the experiments of M. Bréant.

Alloys of Platinum and Bismuth.—These metals

combined produce but brittle alloys, which have only been experimented upon in a scientific manner by Mr. Lewis. With alloys ranging from 1 to 24 parts of bismuth to 1 of platinum, Mr. Lewis has obtained brittle products, nearly as mild as bismuth; their fracture had a foliated appearance, and, by contact with the air, acquired a purple, violet, or bluish color.

Alloys of Platinum and Antimony.—The combination of platinum with antimony gives a dark gray alloy, hard, harsh, and brittle, whose finely granular fracture is a shade darker than that of either metal. A trace of antimony is sufficient to render platinum brittle.

According to Berthier, when a mixture of 2 equivalents of antimony and 1 of platinum is heated at a high temperature, part of the antimony is volatilized, and the remaining alloy is compact, very brittle, with a lamellar fracture, a great lustre, crystalline, and platinum-gray, at the surface, but darker than antimony.

Alloys of Platinum and Nickel.—We have no data on these alloys, which do not seem to have been examined by chemists.

Alloys of Platinum and Arsenic.—These metals appear to combine in all proportions. A very small proportion of arsenic is sufficient to make platinum brittle. According to Thénard, an alloy of 20 parts of arsenic and 2 of platinum presents the following characteristics: a grayish-white color, great brittleness, and fusibility below a red heat. The air at the ordinary temperature has no action on this compound.

In such alloys the arsenic becomes separated by a high temperature, and leaves the platinum in a spongy state.

Amalgams of Platinum.—These amalgams are very difficult to produce. Mercury has no action even upon forged or drawn out platinum. However, with the aid of heat we may obtain platinum amalgams of a fine silver-white color, and which may be kept without tarnishing. These amalgams, which are soft at the beginning, gradually become hard and brittle. They are decomposed by heat, and are generally formed of mercury 73 parts and platinum 27.

GENERAL OBSERVATIONS.—The alloys of platinum, most of which present no practical interest, have been especially the subject of scientific studies and of laboratory experiments. Their principal applications have been the construction of reflectors, the manufacture of weapons, and certain precious alloys with gold, silver, or copper.

The high price of platinum causes it to be rarely used. Moreover, the high temperature necessary to fuse it prevents or renders very difficult the practical production of those of its alloys which might become useful. Most of the applications of platinum belong to the chemical arts, where this metal is employed for crucibles, capsules, retorts, &c. On account of its comparative infusibility or unalterability, platinum has been very useful for insuring the success of a great many delicate operations, which require that the vessels employed should not suffer any alteration capable of exerting a detrimental influence on the results.

Aluminium and its Alloys.—We shall not follow for these alloys the order adopted with the preceding metals. Aluminium is a comparatively new metal, the combinations of which with the other metals have as yet been little experimented upon up to the present day. Its most important industrial applications have been its alloys with copper.

Before aluminium had been obtained in the metallic state, its combinations, like those of the metals of the next chapter, had presented just enough interest to attract the attention of science. Alloys were not then thought of, and the experimenters were trying chemical assimilations, rather than mechanical alloys, in the full sense of the word.

From examination of works treating on the question of alloys, we find very few data concerning the introduction of alumina, not aluminium, into other metallic compounds.

Alumina, which, in the natural state, is combined with a certain number of metals, especially with iron, takes from their useful qualities, when it remains in combination after fusion, rather than imparts new ones to them.

Taking for granted that the damaskeened appearance of wootz or Indian steel, after being forged and polished, was due to alumina, several learned persons have studied the alloys of steel and alumina.

Small pieces of steel were submitted to a protracted and very high temperature, and the resulting carbides having been powdered and mixed with pure alumina, after a powerful heating in a crucible, gave a white alloy, very brittle, and granular in structure.

From 50 to 70 parts of this alloy, melted with from 500 to 700 parts of good steel, gave a metal which, after having been forged, polished, and treated by diluted sulphuric acid, had the damaskeened appearance of wootz steel.

The specific gravity of this compound, not hammered, was 7.665. Although presenting a lamellar fracture, the metal was sufficiently malleable to be drawn out without flaws or cracks. The grain, after the hardening process, was exceedingly fine and hard.

We may be permitted to think that these results of more or less authenticated experiments require to be confirmed by new experiments, more thorough and complete. So many savans have claimed that traces of various metals, added to steel, would improve and transform the properties of this metal, without the results bringing an entire certitude, that we should desire new experiments on the subject; the previous experiments having been confined to the laboratory, and the results being due to fortuitous circumstances which could not be repeated in daily practice.

There is no doubt that, since industry produces metallic aluminium, it will be possible to combine this metal in all proportions with the majority of the known metals. And supposing that, instead of true alloys, we obtain only mixtures, these, by their metallic nature, will find more or less important or useful applications in the arts, and, at all events, will furnish more correct data than those in the possession of science up to the present day.

Metallic aluminium, as extracted from alumina, the base of clays and kaolins, so abundant in nature, approaches iron, cobalt, chromium, and nickel in its chemical properties; and gold, silver, copper, tin, zinc, &c., in its physical properties. Its specific gravity, however, is an exception among metals, and is as low as 2.60, while the average specific gravity of the known metals reaches 7.20

Aluminium was, for the first time, half a century ago, isolated from alumina by Wöhler, a German chemist; but this metal exhibited its true characteristics, only fourteen years ago, through the experiments of M. Sainte-Claire Deville.

The applications of aluminium were quite exaggerated at the beginning. Being light, easily laminated, embossed, drawn out, and chased, it was welcomed by fashion with a favor too great to be lasting.

At the present day, we have passed through that infatuation for a metal, the qualities of which are not good enough to rank it among the precious metals. Dull in color and soft, aluminium was too expensive to vulgarize its applications. Although it has been extolled too much, this metal is, nevertheless, a very interesting conquest of modern metallurgy. Its manufacture and uses, although limited, have been useful in the arts; and the works at Nanterre, where aluminium is produced, refined, and worked, are to be ranked among those manufacturing establishments which have recourse to the laboratory only for improvements and new results. Other works, in England and Germany, have been created in the same manner as those of Nanterre, and have for several years already produced aluminium and its alloys on a manufacturing scale.

We may form several alloys with aluminium and tin, zinc, silver, iron, platinum, copper, &c.; but most of them have no practical interest. Alloyed with the precious metals, aluminium takes from them a part of their intrinsic value, without imparting new qualities to them as a compensation. Combined with certain industrial metals, such as zinc, tin, iron, &c., it loses itself part of its intrinsic value, without acquiring, from what we actually know, any peculiar property which might widen the field of its useful applications.

Its alloys with copper are the only combinations which, at the present time, have been seriously practised.

To speak exactly, the combinations of aluminium with the other metals are rather associations than true alloys. The low specific gravity of aluminium is a drawback to an alloy easily made by the direct process. We are obliged to introduce aluminium, gradually and by small portions at a time, into the other melted metals, in order to saturate them, rather than to produce a combination.

For the compounds of copper and aluminium, the best kinds of refined copper are necessary.

The atomic proportions appear to range between 10 parts of aluminium and 90 of copper.

When an ingot of aluminium is introduced into the middle of a bath of molten copper, the latter is immediately cooled off, and becomes hard. It is only after a vigorous and continuous stirring that the nearly coagulated mass becomes fluid again, and the combination takes place. According to Mr. Morin, the director of the manufactory of Nanterre, very homogeneous alloys are obtained with the proportions of 5, 7½, and 10 per cent. of aluminium; whilst with the proportions of 6, 7, or 8 per cent. there is no thorough mixture or combination. The alloys with 5 and 10 per cent. of aluminium are both of a golden-yellow color, whereas that with 7½ per cent. gives a metal having a greenish tint, perfectly different from that of the two other compounds. We may be allowed to suppose that, in such cases, there is some peculiar process of handicraft which cannot be seen by the observer; and that the above indications would require to be confirmed by other experiments.

The direct mixture, by first fusion, of 10 parts of aluminium and 90 of copper, gives a brittle metal, which increases in strength and tenacity only after several successive fusions. At each operation, a little aluminium is lost.

However, when the compound has been remelted three or four times, the proportion of aluminium does not seem to change, and the alloy may be remelted several times without alteration. These fusions are effected in crucibles. The *aluminium bronze*, when melted several times, is homogeneous, and possesses sufficient expansion to fill the remotest parts of the moulds. It may be cast into very thin and sharp objects, nearly as well as good statuary bronze. On the other hand, when the pieces are bulky, this aluminium bronze is subject to shrinkage, and requires numerous runners and a heavy feeding head (*dead head*).

Aluminium bronze may be forged at a brown-red heat, and hammered until cooled off, without presenting any flaw or cracks. This alloy, the same as copper, is rendered milder and more ductile by being plunged into cold water when hot.

The specific gravities of the alloys of copper and

aluminium are sensibly proportional to the amount of aluminium. According to Messrs. Bell Brothers, of Newcastle, the specific gravities of compounds of copper and aluminium are:—

For an alloy of 3 per cent. of aluminium . .	8.691
" " 4 " " " . .	8.621
" " 5 " " " . .	8.369
" " 10 " " " . .	7.689

From the experiments made by Colonel Strange, and stated in the Proceedings of the Royal Astronomical Society of London, it results that:—

The resistance to traction of aluminium bronze is 5328 kilogrammes per square centimetre; whereas that of the ordinary ordnance metal (bronze) of Woolwich is 2552 kilogrammes.

The resistance to compression is feeble; the metal becomes flattened under the charge, the same as with soft metals.

The malleability is great, although no figures accompany the experiments. Aluminium bronze may be forged with great facility. From a dark red heat up to a limit near its point of fusion, this metal behaves perfectly well under the hammer.

The absolute rigidity was not determined. Mr. Strange's experiments were confined to the relative rigidity of brass, ordinary bronze, and aluminium bronze; and the results were that aluminium bronze was about forty times as rigid as brass, and three times as much so as ordinary bronze.

Other experiments have shown that aluminium bronze does not expand or contract so much as ordinary bronze, and does so, much less than brass. That under the tool aluminium bronze produces long and resisting chips, does not clog the file, &c. That, in the melted state, this metal expands very much, and is fit for the sharpest castings; but that, as it cools off rapidly, it is subject to shrinkage, and hence to cracks. At last, that although not being entirely inoxidable, it is, however, not so easily tarnished by contact with the air as polished brass, bronze, iron, steel, &c.

At all events, notwithstanding its imperfectly observed qualities, it is certain that aluminium bronze has not yet found a large place in the arts. The price of this alloy, which ranges from 15 to 50 francs ($3 to $10) per kilogramme (2.20 pounds), whether in the raw state or more or less worked, and we do not speak of artistic works, is certainly an impediment to its common use. If we add that when polished its color is not very pleasing, and does not, whether by its tint or lustre, resemble those of the precious metals; that its unalterability is not entirely demonstrated; we will understand the slowness of the progress of aluminium bronze in public favor.

The articles actually manufactured from aluminium bronze are generally copies of goldsmith's ware. Spoons, forks, dessert-knives, supports for decanters, coffee-pots, &c., are made with the alloy holding the maximum of aluminium. Candlesticks, small jewelry ware, broaches, buckles, the accessory parts for surgical or mathematical instruments, etc., are made with an alloy of a lower grade.

In fact, the tendency is to substitute these alloys for many gilt, silvered, or plated articles, when on account of their peculiar properties they may present the same advantages of duration at nearly the same cost.

IV.

ALLOYS OF THE METALS RARELY OR NEVER USED IN THE ARTS.

We shall include in this chapter the mixtures, rather than alloys, formed between themselves or with the preceding metals, by certain metals which chemistry classes among metalloids,* rather than among metals proper.

Most of the elementary bodies which we now have to examine are rare, little known, scarcely or never used, and belong more to science than to the arts. Several of them have not been obtained in the metallic state; and we believe that under such conditions their combinations are more interesting from a scientific point of view than adapted to use in the arts; more for the laboratory than for the foundry; and therefore not within the limits of this work.

This chapter shall be short, and limited to concise indications relating to alloys.

However, brief as are the indications we have to give concerning the more or less useful combinations of the following metals:—

Manganese,
Chromium,
Cobalt,
Cadmium,
Titanium,
Uranium,
Tungsten or wolfram,
Molybdenum,
Osmium,
Iridium,
Palladium,
Rhodium,
Tellurium,
Silicon or silicium,
Potassium,
Sodium—

* The term *Metalloid* is applied in chemistry to those elementary bodies which, combined with oxygen or hydrogen, may act as acids, and whose oxides do not play the part of bases. Silicium is the only metalloid among the metals to be examined in this chapter.—*Trans.*

we shall precede them by a rapid sketch of the history and characteristics of these metals. We shall not mention any of those which are scarcely known by chemists themselves.

MANGANESE.

Discovered in the metallic state, in 1774, by Scheele and Gahn. Specific gravity about 7.05. Fascicular and crystalline fracture, of a grayish-white color, resembling that of white pig-iron. Less fusible than cast iron. Without smell or savor. Brittle and difficult to file. According to Mr. Regnault, however, it possesses a certain ductility and malleability which would approach that of iron, if it could be obtained in a pure state.

In order to preserve manganese, it must be kept from contact with the air. This metal has a great tendency to become oxidized, and its surface is rapidly covered with a dark brown oxide as soon as it is exposed to a damp atmosphere.

Manganese, according to Bergmann, unites with copper and gives a very malleable alloy, of red color, which after some time turns to a greenish-brown.

According to Berthier, alloys of copper and manganese are ductile, and each metal possesses a great affinity for the other.

The same savant has tried the following alloys, made by heating the mixture in a brasqued crucible, and obtained in the shape of metallic buttons.

	1	2	3	4
Protoxide of manganese	4.46	8.92	8.92	17.84
Metallic copper	31.64	31.64	15.82	15.82
Charcoal	0.50	1.00	1.00	2.00
Borax	0.50	1.00	1.00	1.00
	37.10	42.56	26.74	36.66

Alloy No. 1 gave a compact metal of a grayish-white color, shaded with red, perfectly ductile, very tenacious, and with a granular and scaly fracture. The proportion of manganese was about 10 per cent.

Alloy No. 2 was platinum-gray, ductile, tenacious, and susceptible of a fine polish.

Alloy No. 3 gave similar results to the preceding ones, although its composition after fusion was 2 atoms of copper to 1 of manganese. The composition of alloy No. 2 was 4 atoms of copper to 1 of manganese.

Alloy No. 4 gave a well-melted metal, iron-gray, ductile, very tenacious, susceptible of acquiring a very fine polish, and with a scaly and at the same time fibrous fracture. This metal, the composition of which was about 4 atoms of copper to 3 of manganese, exhaled a smell of hydrogen when breathed upon.

The composition of these alloys shows a great affinity between the two metals; because, without the presence of the copper the proportion of reduced manganese would not have been so considerable.

Gold, like copper, may be alloyed with manganese. This latter metal, melted with 33 per cent. of gold, forms a hard alloy of a light gray color, with little ductility, and having a granular fracture. With only 10 per cent. of gold the alloy becomes entirely ductile, finely granular, and of a light gray.

The alloy composed of 12 per cent. of manganese and 88 per cent. of gold is a pale yellowish-gray, with a fine lustre, similar to that of polished steel. This alloy, which is less fusible than gold, is very hard and slightly ductile. Its fracture presents a spongy appearance, the grains are coarse, and the color is a reddish-gray. It is not altered by contact with the air. According to M. Dumas, the above proportion of manganese is the maximum which can be employed without debasing the gold too much.

Manganese is often found combined with certain kinds of pig-iron. But these forced combinations are to be found in white, lamellar, and very brittle pig-iron, and there seems to be no advantage in direct alloys of iron with manganese.

It appears, however, that the presence of manganese in pig-iron is valuable for the manufacture of steel. In this respect it would be interesting to study more thoroughly than has been done what is the action of manganese on pig-iron. The main point would be to obtain a combination of manganese with pure pig-iron—that is, deprived of such other substances as are susceptible of altering its qualities. Several kinds of pig-iron, holding manganese, have been found by analysis to contain also copper, zinc, silica, alumina, phosphorus, &c. Most of these substances being prejudicial to pig-iron, whether for casting or the manufacture of iron and steel, it is certain that all of the bad effects are not attributable to manganese.

Berthier has indicated an alloy of—

Copper	0.661
Tungsten	0.216
Iron	0.091
Manganese	0.031
	0.999

which is semi-ductile, very hard, susceptible of a fine polish, and nearly as red as pure copper. This skilful chemist has thought that, by increasing the proportion of copper, the alloy would become entirely malleable. as fine as copper, harder, and a great deal less fusible.

This would be a curious experiment to make, unless it has already been done. But, as regards alloys, we must be prepared for unforeseen results; and changes in the proportions will not always be accompanied by corresponding transformations in the nature of the alloys, such as would have been presupposed from the composition of the primitive alloy.

From what we know, no other experiments on

alloys of manganese have been made. At least, none have been published.

CHROMIUM.

Discovered by Vauquelin, in 1797. Specific gravity =5.9. Very hard and brittle. Scratches glass, and is very slightly fusible. Its color resembles that of tin, and, after polishing, acquires a fine metallic lustre.

Chromium, in the natural state, is found combined with iron and lead, forming chrome iron, and chromate of lead or *crocoïde*. It is difficult to obtain in an entirely metallic state; by the known processes, it is produced as an agglutinated grayish mass, or a dark gray powder. In either case, it is not completely pure. Chromium is not very oxidizable by contact with the air, at the ordinary temperature; but, at a red heat, and with that contact, it becomes incandescent by the absorption of oxygen, and is changed into the green oxide of chromium.

The chemical combinations of chromium are remarkable for their colors.

Experimenters appear to have studied only the combinations of chromium with iron and steel. From their researches it has been ascertained that chromium has a powerful affinity for iron, and may be alloyed with this metal in all proportions.

According to Berthier, if we submit to a powerful heat, in a brasqued crucible, a mixture of the oxides of chromium and iron, they are perfectly reduced, and we may obtain, in all proportions, intimate and homogeneous combinations of the two metals.

These alloys are generally hard, brittle, crystalline, grayish-white, and, when polished, more lustrous than iron. With an increase in the proportion of chromium, they become proportionally more refractory, less magnetic, and more indifferent to the action of acids.

The alloy made of—

Iron	68.60
Chromium	31.40
	100.00

has a fibrous structure, a white color nearly like that of silver, and is very brittle and difficult to file.

The alloys of chromium and iron have not, as yet, been used on a very large scale in the arts. In case they should, it would be better, according to Berthier, to substitute, in the mixtures, the chrome ore (chrome iron) for the pure oxide of chromium. The chrome ores are not scarce, and a large deposit has been found in the department of Var (France).

In his experiments on the combinations of chromium with iron and steel, Berthier has employed the alloys of chromium and iron for introducing the former metal into cast steel.

The alloys of steel and chromium made by that process, and holding from 1 to 2 per cent. of chromium, gave a metal which, like *wootz* or *Indian steel*, could be polished, and then damaskeened by means of diluted sulphuric acid. The damaskeened pattern (the white portions of which were chromium, upon which diluted acid has no action) presented variegated veins, with a brilliant silver lustre, and similar to those obtained in the alloy of silver with steel.

Several manufacturers of arms in Belgium have, by similar processes, tried the alloys of steel and chromium for their damaskeened blades. We believe that these alloys are in actual use, but that steel of cementation has been substituted for cast steel, which was employed in the experiments.

Other trials made by Berthier on alloys of chromium and copper, chromium and tin, do not appear to have been applied in the arts. We shall detail them as subjects of information, reminding our readers that among these brief data they may find a basis for new studies, which, if made in a practical manner, may possibly lead to unforeseen results.

According to Berthier, the alloy made of—

Copper	0.912
Chromium	0.088
	1.000

is malleable and harder than copper. It has the same color as the latter metal, and will acquire a fine polish.

The alloy composed of—

Tin	0.808
Chromium	0.192
	1.000

is grayish-white, soft, semi-ductile, harder than tin, but cannot be laminated. Its fracture is granular, and iron-gray.

Cobalt

Was discovered by Brandt in 1733. Specific gravity 8.6. Fracture, a reddish-gray; when polished, its color is of a steel-gray, as magnetic as iron, more fusible, and less ductile. It takes a fine polish. Its tenacity is remarkable, and, according to M. Regnault, nearly double that of iron.

In the natural state, cobalt is found combined principally with sulphur and arsenic, under the names of arsenical cobalt or *smaltine*, and gray cobalt or *cobaltine*.

Pure cobalt, or its alloys, have no industrial uses. Its oxide is employed for the manufacture of azure blues, Thénard blue, &c., for the enamels of decorators on china and glass ware. The savans of this period have paid considerable attention to the chemical combinations of this metal, which appear to have been employed, from the earliest ages, by the Egyptians, Greeks, and Romans, for their glasses and blue enamels.

Cobalt is not so much affected by dampness as iron. However, by the permanent action of damp air, it becomes covered with a pellicle of a fine black oxide. Heated in the air, it is transformed into oxide.

Berthier has tried an alloy made of—

Copper	68.2
Cobalt	31.8
	100.0

the composition of which was ascertained after fusion. This alloy was compact, ductile, tenacious, of a white slightly tinged with red, strongly magnetic, and susceptible of a fine polish.

An alloy of—

Tin	80
Cobalt	20
	100

was very fusible, easily cut and hammered, although brittle, and with a rugged and crystalline fracture.

These laboratory experiments, made in brasqued crucibles, and investigated upon buttons weighing no more than 15 to 20 grammes, may give some indications of the mode of operation, but do not actually present any practical result of interest in the arts.

Indeed, what is to be expected from such alloys, which, as all those we are now examining, are at present incapable of furnishing economical compounds? Only unforeseen results, which may be applied to the arts, in cases where the alloys and metals now in use do not possess the qualities desired. Then, a few experiments, made in the manner of Berthier, will be sufficient to show the investigator if he is moving in the right direction.

Cobalt has also been tried with iron. Berthier says that such alloys possess the same qualities as pure iron, and are of a whiter color.

CADMIUM.

Discovered by Stromeyer and Hermann in Germany, about the same time, in 1818. Its color is white, with a tinge greener than that of tin. Possesses as much lustre as tin. Fracture, fibrous and crystallizing in regular octahedrons. Specific gravity, 8.6. Fusible below a red heat, and volatilizes at about 400° C. Malleable, ductile, somewhat harder than tin, and may be laminated and drawn out.

Cadmium is found in the natural state, combined with sulphur and zinc in several varieties of calamine and blende. It is not sensibly oxidized at the ordinary temperature, but, when heated to redness, it volatilizes sooner than zinc, and its vapors burn with brilliancy. Distilled in a retort, pure cadmium may be obtained in the shape of regular and crystalline drops.

The great facility with which cadmium volatilizes has been the serious drawback to the formation of its alloys and their study.

Cadmium is very easily dissolved in mercury, even at the ordinary temperature. The amalgam is of a very fine silver-white, and its texture is granular and crystalline. It melts at 75° C., and when cooled off is hard and very brittle. Its specific gravity is above that of mercury.

TITANIUM.

Its oxides appear to have been studied from 1790 to 1795 by Gregor and Klaproth. Since 1821, its combinations have been investigated by the chemist Rose.

Combined with various substances, especially with iron and oxygen, carbon and nitrogen, titanium is one of the most refractory of metals. Reduced to the metallic state, it forms a black and amorphous powder, similar to that of iron reduced by hydrogen at a low temperature.

Heated in contact with the air, titanium burns and produces a vivid scintillation; the incandescence is sudden, and the metal is projected out of the crucible, when it is heated in contact with the oxides of lead or copper.

We must acknowledge that the black powder of reduced titanium is far from presenting a characteristic metallic appearance. Hence a great difficulty of assimilation, which has prevented experimenters from trying the alloys of titanium. The only experiments known were based on the alloys of this metal with iron, and the results have been negative. Karsten, in his work on the metallurgy of iron, mentions an attempt to combine titanium with steel; and although the proportion of titanium was only 1 per cent., the alloy did not take place, and the titanium was found irregularly scattered throughout the mass.

Uranium

Was isolated from the oxide known under that name, by M. Peligot, in 1842. Metallic uranium, whether in a black powder or aggregated in the shape of small laminæ, presents a lustre similar to that of silver. In the latter case, it appears to possess a certain malleability.

This metal, heated to a temperature above 200° C. in presence of the air, burns with much brilliancy, and is transformed into a dark green oxide. At the ordinary temperature it does not decompose water, and is not altered by the contact of the air.

With acids, the protoxide of uranium, UO, produces green salts; the sesquioxide, U^2O^3, gives yellow salts. The latter oxide is employed for imparting to glassware a yellow shade with a green tinge.

It has not been possible, up to the present time, to combine uranium with the other metals. This is most likely due to its imperfect metallic state, which, like that of titanium and certain other metals obtained in the form of powder, is not adapted to the production of alloys.

TUNGSTEN

Was isolated from wolfram, in 1790, by the brothers d'Elhuyart. We obtain it, either as a black powder or a solid mass, rather coagulated than melted, which acquires under the file a certain metallic lustre, of a dull gray color.

This metal is very expensive, and but slightly fusible. Its specific gravity is considerable, and attains 17.6.

In the natural state, tungsten is combined with lime or lead, forming the *schéelite* or *schéelitine*; and with iron and manganese in *wolfram*.

During the last few years, the alloys of tungsten with cast iron, steel, and wrought iron have attracted a great deal of attention, in the hope that these metals would acquire new qualities of resistance and hardness. Mr. Leguen, major of artillery, has superintended all the experiments made in this direction, for improving the quality of the metals employed in the manufacture of ordnance, and other weapons.

We do not think, whatever has been said, that these experiments, up to the present time, have given conclusive results. We shall, however, relate here, for the instruction of our readers, the principal data of the report of Mr. Leguen to the minister of war.

A small proportion of wolfram imparts to cast iron an extraordinary hardness and tenacity. The latter quality increases in a greater ratio than the former, as the proportion of wolfram also increases up to a certain limit. Therefore, it is important to vary the proportion of wolfram according to the future uses of the cast iron employed. This proportion may vary from ½ to 5 per cent. of wolfram. The wolfram employed in the experiments of Mr. Leguen, was extracted from the mine of Puy-les-Vignes, near Saint Léonard, in the Haute Vienne. It is the only mine of this kind known in France. The wolfram, imbedded in a very hard gangue of quartz, contains about 60 per cent. of tungsten, the remainder being iron, manganese, and oxygen.

Mr. Leguen infers from this composition that tungsten, being in a proportion at least three times that of the other two metals together, will perform the principal part in the modifications imparted to cast iron by this ore. He explains, furthermore, that the small proportion of manganese introduced into the alloy does not sensibly act upon it; that the addition of iron has no other effect than to increase the bulk of the cast iron; and that, therefore, the increase in hardness and resistance is due to tungsten alone.

If we examine the question by the light of the experiments of Mr. Stirling, in England, which tend to show that the tenacity of cast iron is considerably increased by the addition of wrought iron; if we also state that many persons believe that manganese in cast iron imparts to it a greater resistance—we may well have some doubt whether tungsten alone, as Mr. Leguen says, is the true cause of the increase of resistance of cast iron, with which wolfram has been alloyed.

We have ourselves, with the prepared and fritted wolfram, sent to us by the owner of the mine at Puy-les-Vignes, made careful experiments on the introduction of wolfram into cast iron; and these experiments repeated several times gave us samples, which being tried, gave results sometimes favorable, sometimes unfavorable, to the action of wolfram. The figures obtained by these trials exhibited such slight differences, that it would be as proper to suppose these differences due to possible variations in the nature of the cast iron, from one smelting to another, as to the presence or absence of wolfram.

It is well known that metals in general, and pig-iron especially, may widely differ in their resistance, even when they have been uniformly mixed, melted, cooled, &c. We have also demonstrated in another work, that four railroad chairs, cast at the same time in the same mould, presented in certain cases differences quite considerable in their resistance. Therefore, we may infer, *a fortiori*, that these differences will take place if trial bars are cast at different times, at variable temperatures, and although the whole operation appears to be conducted in the ordinary regular manner, with the same qualities and proportions of metal for the mixtures.

Therefore, we should consider it quite natural that results from certain trials have caused wolfram to be regarded as possessing the qualities necessary for considerably increasing the resistance of cast iron.

Certain bars, tried by the skilful directors of the Conservatoire des Arts et Métiers, have indicated that wolfram improved cast iron, but it was not ascertained whether the bars with wolfram, and those without, had been cast on the same day; or, notwithstanding the precautions taken to operate in exactly the same conditions, if bars of cast iron without wolfram, and cast at different times, would not have presented the same differences.

It will be sufficient, in order to a better understanding, to state the results of several experiments made by ourselves, at the Marquise iron-works, in 1862.*

* The results are shown by figures indicating the relative resistance. In the trials by shock, the square bars had their sides equal to 4 centimetres, and were put upon edged supports, 16 cen-

A. Gray cast iron, from Marquise, and without any admixture. Six bars tried by shock:—

1st bar, breaks at	. . .	0.65 metre of fall.
2d " " "	. . .	0.75 " "
3d " " "	. . .	0.70 " "
4th " " "	. . .	0.80 " "
5th " " "	. . .	0.85 " "
6th " " "	. . .	0.90 " "
Average	. .	0.775

B. The same cast iron, with ½ per cent. of wolfram. Six bars tried by shock:—

1st bar, breaks at	. . .	0.55 metre of fall.
2d " " "	. . .	0.55 " "
3d " " "	. . .	0.60 " "
4th " " "	. . .	0.65 " "
5th " " "	. . .	0.75 " "
6th " " "	. . .	0.85 " "
Average	. .	0.658

C. The same cast iron, with 1 per cent. of wolfram. Six bars tried by shock:—

1st bar, breaks at	. . .	0.75 metre of fall.
2d " " "	. . .	0.80 " "
3d " " "	. . .	0.90 " "
4th " " "	. . .	0.90 " "
5th " " "	. . .	0.90 " "
6th " " "	. . .	0.95 " "
Average	. .	0.866

D. The same cast iron, with 8 per cent. of iron turning scraps, and without wolfram. Six bars tried by shock:—

timetres distant from centre to centre. The shock was given by a ball weighing 12 kilogrammes.

In the trials by flexion, the numbers indicate the breaking strain of square bars (side = 25 millimetres), put upon edged supports 50 centimetres distant from centre to centre.

1st bar, breaks at	. . .	0.80 metre of fall.
2d " " "	. . .	0.80 " "
3d " " "	. . .	0.80 " "
4th " " "	. . .	0.85 " "
5th " " "	. . .	0.85 " "
6th " " "	. . .	0.85 " "
Average	. .	0.825

A′. Repetition of the experiment A. Six bars tried by shock:—

1st bar, breaks at	. . .	0.65 metre of fall.
2d " " "	. . .	0.65 " "
3d " " "	. . .	0.70 " "
4th " " "	. . .	0.70 " "
5th " " "	. . .	0.70 " "
6th " " "	. . .	0.75 " "
Average	. .	0.692

B′. Repetition of the experiment B. Six bars tried by shock:—

1st bar, breaks at	. . .	0.70 metre of fall.
2d " " "	. . .	0.70 " "
3d " " "	. . .	0.70 " "
4th " " "	. . .	0.75 " "
5th " " "	. . .	0.75 " "
6th " " "	. . .	0.75 " "
Average	. .	0.725

E. Gray cast iron of Marquise, the same which had been employed in the previous experiments. Six bars tried by flexion:—

1st bar, breaks by a strain of	.	2900 kilogrammes.
2d " " " "	.	2900 "
3d " " " "	.	3000 "
4th " " " "	.	3000 "
5th " " " "	.	3300 "
6th " " " "	.	3300 "
Average	.	3066

F. The same cast iron, with ½ per cent. of wolfram. Six bars tried by flexion:—

1st bar, breaks by a strain of	.	2700 kilogrammes.
2d " " " "	.	3000 "
3d " " " "	.	3000 "
4th " " " "	.	3000 "
5th " " " "	.	3000 "
6th " " " "	.	3100 "
Average	.	2966

G. The same cast iron, with 1 per cent. of wolfram. Six bars tried by flexion:—

1st bar, breaks by a strain of	.	2600 kilogrammes.
2d " " " "	.	2700 "
3d " " " "	.	2700 "
4th " " " "	.	2900 "
5th " " " "	.	3100 "
6th " " " "	.	3100 "
Average	.	2850

H. The same cast iron, with 8 per cent. of iron turnings. Six bars tried by flexion:—

1st bar, breaks by a strain of	.	2700 kilogrammes.
2d " " " "	.	2700 "
3d " " " "	.	2700 "
4th " " " "	.	2900 "
5th " " " "	.	2900 "
6th " " " "	.	3100 "
Average	.	2833

The examination of these results shows that cast iron without wolfram, and cast iron with wolfram, give, excepting the results of the trials *B*, figures sufficiently near to suppose that the differences are due to the anomalies presented by the same cast iron in similar experiments. The influence of wolfram is not sufficiently demonstrated, even in the experiments *C*, where it is the most perceptible, to be admitted without dispute. Moreover, the trials *D*, where iron had been added to cast iron, gave results so near those of *C*, that we cannot say whether it is the iron or the wolfram which has increased the resistance of the metal.

On the other hand, the trials *A′* and *B′*, repeated under exactly the same conditions as those of *A* and *B*, come into direct opposition to the former trials, and show that wolfram has a beneficial influence, while in the former cases it was rather hurtful.

It may be objected that the preparation of the alloys has possibly been defective. Indeed, it is difficult to melt wolfram, which, when pure, is nearly infusible.

The nature of the elements of cast iron appears to facilitate its fusion; nevertheless, the alloy is difficult, on account of the great specific gravity of tungsten. But we are certain that we took all the necessary precautions to obtain the mixture, whether operating in a crucible or in a cupola.

The experiments of Mr. Leguen were conducted in a similar manner, as regards the fusion in crucibles. The cast iron and the wolfram were charged at the same time in the red-hot crucibles, and the temperature was raised afterwards. The trials were, like ours, made upon square bars (side = 0.04 metre), first with cast iron only, and then with the same metal combined with 1, 1½, 2, and 2½ per cent. of wolfram. The result of the trials has shown an increase of tenacity by each addition of wolfram, but not in proportion to the quantities employed. However, the ratio of increase of tenacity appears to have been regular up to 2½ per cent. of wolfram.

Mr. Leguen infers from his experiments that as cast iron may have its tenacity increased one-third by alloying with wolfram, all ordnance should be transformed on these new bases. This conclusion goes too far, the more so as Mr. Leguen recognizes himself that the trials have been insufficient, and should be repeated in various ways.

From cast iron Mr. Leguen passes to steel, which, according to the same authority, is even a great deal more improved by wolfram. Steel combined with wolfram ought to acquire similar qualities to those of steel combined with pure tungsten, or with molybdenum, chromium, titanium, and alumina, which substances, according to certain experimenters, may form five damaskeened compounds. According to Mr. Leguen, careful experiments on a practical scale have been attempted in order to impart, by means of wolfram, various degrees of hardness and tenacity to the steel intended for the manufacture of files, cutting instruments, weapons, &c. But, at the present day, we cannot say that anything in that line has been introduced into the art. On the contrary, we know that an important steel-works, which had great faith in the alloys of wolfram and steel, has abandoned the idea, after a few experiments, which demonstrated the difficulty of arriving at certain and unfailing results.

Consequently, it seems better to wait before forming an opinion on the influence of wolfram upon steel or cast iron. Wootz, the Damascus steel of the East, and the other compounds where steel appears with peculiar properties, are rather natural products than alloys proper, and, therefore, cannot well be compared with the alloys which we are studying.

Mr. Leguen also considers the alloys of wolfram with copper and tin, in order to improve the bronze for ordnance. These alloys are exceedingly difficult to obtain, on account of the differences in fusibility and specific gravity of wolfram, and of the component metals of bronze. The alloy of wolfram and copper is very difficult, and there, as with cast iron, nothing demonstrates that wolfram increases the hardness or tenacity of copper. Our own experiments gave no useful data, and too often, after running out the copper or the brass, we found the wolfram in a pulverulent state, uncombined with the copper and tin, notwithstanding all the precautions taken by the founder for rapidly melting, stirring, and running out.

To sum up, we will say that in our opinion, and that of many of the chemists who have studied the action of wolfram, if tungsten could be separated from wolfram in an economical way, it might give more important and more conclusive alloys.*

MOLYBDENUM.

Obtained by Scheele in 1778, and isolated afterwards by Hielm. Specific gravity, 8.6. Color, a dead white, susceptible of a fine polish. Is found in the natural state combined with sulphur or lead. It is obtained as a grayish powder, which acquires a metallic lustre by being burnished, and sometimes in the shape of small melted masses which resemble unpolished silver.

Molybdenum is easily oxidized. Heated in the presence of the air, it becomes incandescent, and is transformed into molybdic acid.

Molybdenum is without application in the arts. Its combinations with tin have been experimented upon, and Berthier says that the alloy of:—

Tin	83
Molybdenum	7 (or 17?)

is as white, ductile, and tenacious as tin, and may be laminated to thin sheets. Muriatic acid dissolves the tin of the alloy, and leaves molybdenum in the metallic state.

* Mr. C. W. Siemens says that tungsten has the remarkable effect upon steel of increasing its power to retain magnetism when hardened. A horseshoe magnet of tungsten steel has been made which supports twenty times its weight.—*Trans.*

An alloy of molybdenum with lead whitens the color of lead, if the proportion of molybdenum is not over a twentieth; above that, lead becomes harder and darker. Molybdenum unites with certain other metals only in definite proportions, but these alloys present nothing of interest in the arts.

Osmium.

Discovered in 1803, by Tennant, in the ores of platinum, it is generally combined with iridium and ruthenium. Specific gravity, 10. Color, a metallic gray, resembling that of platinum. This metal presents sufficient malleability to be obtained in the shape of aggregated plates, which, however, are easily pulverized by percussion.

It is oxidized by exposure to a damp atmosphere; but, when heated at a low temperature in presence of oxygen gas, it takes fire and is transformed into osmic acid, which volatilizes.

From its chemical properties, Mr. Regnault thinks that osmium should be classified among the metalloids.

Osmium has been tried in an alloy with steel for improving cutting instruments. It is even said that certain steel manufacturers of Sheffield have largely used this metal for their products.

Iridium.

A gray metal found, like the preceding one, in certain ores of platinum. Discovered in 1803 by Tennant and Collet-Descotils, in the black residuum from the treatment of platinum ore with aqua regia. Specific gravity, 15.8.

Iridium is obtained in the shape of a spongy mass, which acquires a metallic lustre by being burnished. It may also be transformed into a very hard and compact mass, which is susceptible of being polished, if the pulverulent metal is wetted, strongly compressed, and then calcined. The specific gravity given above is that of this aggregated and porous metal. Brought to a red heat with potassa or nitre, iridium becomes oxidized and is transformed into iridiate of potassa.

Of course, like those metals which seem to be a universal panacea in developing and improving the qualities of steel, iridium has been combined with that metal, especially by English experimenters.

Messrs. Stodart and Faraday, who have tried iridium on a large scale, claim that this metal produces one of the best combinations with steel, and that the most advantageous proportion for improving the steel for cutting instruments is about 1 per cent. of iridium.

According to Berthier, an alloy of:—

Lead	89
Iridium	11
	100

is whiter than lead, which is rendered harder and more malleable, without any loss of tenacity.

When platinum and iridium can be melted together, which is quite difficult, on account of the refractory nature of the two metals, the resulting alloys are harder than pure platinum and not so easily altered by the action of the fire and reagents. They are, therefore, useful for the fabrication of certain chemical apparatus. We learn from the recent studies of several chemists, that platinum alloyed with one-tenth of iridium has more lustre, is more malleable than pure platinum, and may be hardened. Such an alloy might be useful for metallic mirrors.

We have not seen any other important alloys of iridium, which metal appears to form, with most metals, mixtures rather than complete combinations.

PALLADIUM.

Discovered by Wollaston, in 1803, in certain platinum ores. Specific gravity, 11.5. Unalterable by the air, this metal has a white lustre, slightly duller than that of silver. Very malleable, and may be welded and forged at a white heat. Nearly infusible by the ordinary processes. It is not attacked by certain acids; but hot nitric acid dissolves it readily.

Metallic palladium is actually to be found in the trade, and is a secondary product of certain gold ores, which are a true combination of gold and palladium; such is the *auro-poudre* (gold-powder) of Brazil.

Palladium unites readily with gold, and the alloy is hard, ductile, and platinum-white, when the proportion of palladium is not too considerable. The fracture of this alloy is coarsely granular.

One of the great graduated circles of the observatory of Paris appears to have been made of that alloy, which is dense, hard, and firm enough to receive the finest divisions. M. Regnault states somewhere that this circle is entirely made of palladium. Another author says that the alloy is made of silver and palladium.*

We incline towards the latter alloy, which is easily made, is malleable, ductile, and possesses a fine color, grayer than silver, but whiter than platinum. An alloy of equal parts of palladium and silver has a specific gravity, 11.29.

An alloy of palladium with from 10 to 20 per cent. of silver is employed by dentists for filling teeth.

The ternary alloy of palladium, silver, and gold can be made easily and in all proportions, according to

* In Dana's mineralogy we find that, at the suggestion of Dr. Wollaston, an alloy of palladium—1 part to gold 6 parts—was employed by Troughton for the construction of the graduated part of the mural circle, at the Royal Observatory of Greenwich.—*Trans.*

Berthier. The compounds are ductile, but more dense and elastic than the binary alloys of palladium with silver or gold.

From the preceding indications it seems that in every case palladium, by its white color, its disposition to acquire a fine polish, its resistance to sulphurous fumes and to oxidation, may be successfully employed by manufacturers of philosophical instruments.

Palladium unites more or less easily with certain metals, such as zinc, tin, lead, and platinum. We possess no exact data on these various combinations. Lead, tin, and zinc appear to increase its fusibility, but the compounds remain gray, hard, and brittle. Mr. Fischer has found out that at the moment when the combination of palladium with these metals takes place, the alloy becomes phosphorescent in the crucible.

An alloy of platinum and palladium is harder than platinum, but less ductile. With equal parts of these metals, the compound is gray, possesses nearly the hardness of wrought iron, and has a specific gravity of 15.14.

Palladium may be united with steel, according to Mr. Hervé, author of a work on alloys, from which we borrow a few citations, which we do not endorse, especially when we have not had an opportunity of verifying the results.

The alloy of steel and palladium, with one-tenth of the latter metal, is considered by Messrs. Faraday and Stodart as one of the most useful combinations of steel for instruments which must cut smoothly.

RHODIUM.

Like palladium, rhodium was discovered in platinum ores, by Wollaston, in 1803. Specific gravity, 10.6. Rhodium, so called on account of the pink color of the solution of its salts, is a gray metal, like platinum.

This metal is not oxidized by the air at the ordinary temperature, but when it is in a minute state of division it easily combines with oxygen at a red heat.

Rhodium, like most of the metals of this chapter, is very scarce, expensive, and little known.

According to Wollaston, rhodium is one of the numerous metals destined to improve the qualities of steel. A very small proportion of rhodium ought to render steel much harder and less easily oxidable by a damp atmosphere.

Messrs. Stodart and Faraday, who made at Sheffield numerous experiments for improving steel, found out that the alloys of steel, holding from 1 to 2 per cent. of rhodium, presented very great tenacity, united to such a hardness, that the cutting instruments made with these alloys could bear a tempering heat 30° Fahr. above that of the best Indian wootz, although the tempering point of the latter is 40° above that of the best English cast steel.

A compound of equal parts of steel and rhodium gives, according to the same investigators, a fusible alloy which acquires a magnificent polish, is not tarnished, and therefore very well adapted to the manufacture of metallic mirrors.

Rhodium is not very difficult to alloy with gold, and, if added in small proportion to the latter metal, will increase its hardness without altering its ductility.

Rhodium has not, like platinum and palladium, the property of discoloring gold, therefore it might be used for combining with the latter metal, if rhodium itself were not too scarce and too expensive.

Ruthenium.

Discovered, like the preceding, in platinum ores, but especially in the osmide of iridium, which contains from 5 to 6 per cent. of it. Specific gravity, about 8.6. This metal, which bears a great resemblance to iridium, for which it has often been mistaken, is gray, infusible, does not aggregate by heat, and is scarcely acted upon by aqua regia.

Rutherium is without actual utility, and its alloys are not known.

Tellurium.

Discovered in 1782, by Müller, in a gold ore from Transylvania. It is a bluish-white metal, friable, and with a lamellar fracture. Specific gravity, 6.25.

Tellurium, which possesses much analogy with sulphur in its chemical combinations, is found in the mineral kingdom combined with gold, silver, lead, and bismuth. But it appears to possess the greatest affinity for gold; and for a long time the Transylvania ore, from which Müller obtained tellurium, was known by chemists under the names of *paradoxical gold, problematical gold,* and *white gold.*

No important experiments on the alloys of tellurium with the other metals have been made.

Potassium, Sodium.

We might here examine the possible alloys of certain alkaline and earthy metals. We shall, however, confine ourselves to potassium and sodium.

Potassium, which was discovered in potassa by Davy, is a silver-white metal, with a white lustre, readily tarnished by contact with the air. Its specific gravity is less than that of water, and scarcely attains 0.87. Fusible at 68° C., potassium becomes sufficiently soft to be kneaded between the fingers.

It is nearly as inflammable as phosphorus, and may cause severe burns. In order to avoid its oxidation by the air, it is generally kept in naphtha.

Sodium, also discovered by Davy, presents a great

analogy to potassium. However, it is more tenacious, less volatile, and less fusible. Its specific gravity is about 0.97, and its point of fusion 90° C.

These two metals may be alloyed with the majority of the other metals. But these alloys, or rather compounds, present no great interest for the metallurgic arts; and most of them are decomposed in the presence of air or water.

The various metals we have just examined do not properly belong to the arts.

In order to find real applications for them, it is necessary that they should be obtained at comparatively cheap rates, and that they present the peculiar qualities of tenacity, malleability, and unalterability, so desirable in the arts.

In the form of alloys, their uses would be facilitated by allowing the introduction into common metals of other more rare and expensive metals, which, but for the new qualities they impart, would remain unemployed. This is the reason why we have mentioned a subject where all remains to be studied and applied, as regards their use in the arts.

Therefore, the present chapter is to be considered more as a recapitulation of data and experiments for directing the attention of the experimenter, than as a field already cultivated, in which the crops have only to be gathered. To sum up and to finish the comparison, we open here a new field, where the seeds are few and scattered, and the culture of which is necessary, if we desire, from new and positive results, to arrive at a plentiful harvest.

PART III.

ALLOYS USED IN THE ARTS.

In the last part of this work we shall recapitulate, by distinct industrial categories, the alloys known and adopted in practice.

This classification will allow our readers to ascertain more rapidly, by seeking in the place which they occupy in the arts, the usual metallic compound they require.

By noticing the observations which accompany each kind of alloy, by examining the proportions admitted in practice, and by going back to the various chapters of the first part of this work, which show the characteristic properties of each metal, the possible affinities between various metals, the results obtained by chemists and experimenters, &c., the inquirer will certainly find the bases of new, interesting, or useful combinations.

Among the many alloys employed in the arts, there are certainly several which we have omitted, or incompletely described, or, on the other hand, repeated. The difficulty in a work of this kind lies in the method of classification, and we hope that, considering the order and clearness we have endeavored to introduce into the whole, we shall be forgiven the few omissions or repetitions which have escaped our attention.

I.

BRONZES OF ART.

The component elements of the statuary or artistic bronzes, intended to be gilt, are copper, tin, zinc, and lead combined in various proportions.

We have already described the principal alloys formed by these metals, combined two by two, or by three, or by four. It will therefore be sufficient to sum up in this place the requisite qualities for statuary bronzes, and which are the combinations most generally used in the arts.

The principal conditions required for statuary bronzes, and which we have indicated in our work on foundries, are as follows:—

A yellow-red color, without the yellow green or light yellow shades;

A grain adapted to the work of the file, chisel, and other chasing tools;

Sufficient fusibility and fluidity to fill and reach all the parts of the mould, and reproduce the pattern in all its minutiæ;

An appropriate texture for receiving, without alteration, the mordants imparting the appearance of old bronze (*patine*).

The binary alloys of copper and tin, copper and zinc, rarely fulfil these conditions. The alloys of copper and tin are difficult to produce in one operation, often crack by shrinkage, are not easily chased, and take with difficulty the artificial color of old bronze.

The alloys of copper and zinc are wanting in hardness, and do not resist the action of the chisel sufficiently well. If the proportion of zinc be too considerable, they are but slightly fluid, and do not give sharp castings. If the copper is in too great excess, the surface is full of blow-holes. Moreover, the former are hard and brittle, while the latter are soft and without homogeneousness.

The alloys of copper, tin, and zinc answer best to the wants of statuary, and range between the proportions of:—

Copper 85, zinc 11, tin 5,
Copper 65, zinc 32, tin 3,

which we have already indicated.

However, most of the bronze manufacturers add to these alloys a small proportion of lead, which improves and renders them smooth. With these bases the composition of the alloys remains sensibly within the limits admitted by the brothers Keller, and which are on an average:—

Copper	91.40
Zinc	5.60
Tin	1.60
Lead	1.40
	100.00

These proportions are those of the Column of July, the composition of which was more seriously reasoned out than that of the Column Vendome, whose alloy was composed of:—

Copper	89.35
Tin	10.05
Zinc	0.50
Lead	0.10
	100.00

But, in this case, the proportions were so little attended to, that many pieces, being cast with scarcely any tin, were soft, thick, without relief, and have necessitated considerable expense in repairs and chiselling.

The alloys of several large statues, cast recently, average less copper than those of the brothers Keller. The analyses of the bronze of the statues of Henry

IV., Louis XIV., and Louis XV., cast in Paris, give on an average:—

Copper	82.45
Zinc	10.30
Tin	4.10
Lead	3.15
	100.00

This composition is more economical than that of the Keller bronze, and is well adapted for a statuary bronze.

The ancients, who had no knowledge of zinc, or do not seem to have extracted or worked this metal, employed for their bronzes the ternary alloys, made on an average of:—

With the Romans,

Copper	99
Tin	6
Lead	6
	111

With the Greeks,

Copper	62
Tin	32
Lead	6
	100

However, Roman medals have been found, in which the proportions of copper and zinc were in the ratio of 45 to 1, with a slight addition of lead and tin.

Small bronze statues, found in France at various places where the Roman cohorts had sojourned, also contain zinc. Various bronzes, recently obtained from excavations made at Athens, and of which we had several samples, had an average composition as follows:—

Copper	72
Tin	24
Zinc	2
Lead	4
	102

We must suppose that the ancients accidentally employed zinc combined with lead and tin, but without knowing the characteristics of zinc, the classification of which among the usual metals does not go further back than the sixteenth century.

The manufacture of the bronzes intended for gilding requires fusible and fluid alloys, giving sharp castings, easily chased, cut, and turned, and, besides, possessing such a degree of compactness that the minimum of gold necessary for gilding may be employed.

The alloys of copper and tin are too porous, and too pallid; the alloys of copper and zinc are too pasty, and will absorb too much of the amalgam of gold, with the chance of breaking while cooling after the gilding process. If the proportion of zinc is too considerable, the metal becomes harder, but it loses the yellow color required for gilding.

Therefore, the bronzes for gilding are to be found among the ternary alloys of copper, tin, and zinc; and better yet, in the quaternary alloys of copper, tin, zinc, and lead.

With these bases, according to our personal experience, and the opinion of many experienced founders, the best alloys for gilding are comprised between the following limits:—

Copper	70	82
Zinc	25	18
Tin	2	3
Lead	3	1.5
	100	104.5

These alloys appear to fulfil all the conditions required for the founder, the turner, the mounter, and the gilder.

The experiments related by Darcet in his excellent memoir on the art of gilding bronze, which is still full of interest, although old, confirm these data, and show:—

1. That copper alone is difficult to melt and to cast, is too soft, clogs the file, does not take the gilding well, and requires too much gold;

2. That copper alloyed with zinc in the proportions of 70 to 30, is pasty, soft, not adapted to chasing, but takes the gilding well enough;

3. That copper alloyed with tin in the proportions of 80 to 20, is easily melted and cast, but very dry and brittle under the tools, and too hard to cut. The casting is not sharp, is difficult to scour, and does not take the gold amalgam well.

These defects of the alloys of copper and zinc, and copper and tin, are more or less marked, according to the proportions employed, but they are perceptible, nevertheless, in all the binary alloys of these metals. At the same time, these binary alloys are not well suited to the old process of gilding by amalgam.

This latter inconvenience, it is true, may disappear by the present process of gilding by electricity; but the difficulties of casting, chasing, etc., are not changed, and are sufficient to induce bronze manufacturers to retain the quaternary alloys we have indicated.

In the binary alloys, the compounds of copper and zinc are preferable to those of copper and tin. It is true that the latter are more fluid, but they are too hard and harsh, even with the proportions of tin 10 and copper 90. Their color is too gray, they are polished with difficulty, and resist the action of the burnishing tool.

We shall conclude these indications by giving the composition of the bronzes of various statues, analyzed at the French mint in Paris.

Bronze of the statue of Henri IV., Pont Neuf, 1817.

Copper	89.20
Tin	5.00
Zinc	3.50
Lead	1.20
Iron, loss, &c.	1.10
	100.00

Bronze of the statue of Napoleon, 1833.

Copper	84.80
Tin	5.80
Zinc	6.00
Lead	2.70
Iron, loss, &c.	0.70
	100.00

Bronze of the statue of the Genius of Liberty, Column of July, 1832.

Copper	92.00
Tin	3.00
Zinc	4.20
Lead	0.70
Iron, loss, &c.	0.10
	100.00

Bronze of the statue of J. J. Rousseau, at Geneva.

Copper	85.60
Tin	6.20
Zinc	7.80
Lead	0.40
	100.00

Bronze of the statue of d'Assas, at Vigan.

Copper	91.10
Tin	3.30
Zinc	0.80
Lead	0.60
Iron, loss, &c.	4.20
	100.00

Bronze of the statue of Molière, at Paris.

Copper	90.30
Tin	5.90
Zinc	2.50
Lead	1.20
Iron, loss, &c.	0.10
	100.00

We see that all these alloys correspond to the above quaternary alloys. These compositions are followed in the works of Victor Thiebaut, at Paris, who, at the present time, has quite the monopoly in the casting of large monumental bronzes.

II.

ALLOYS FOR COINAGE.

The conditions which such alloys should fulfil are:—

A perfect regularity in the composition of the alloys.

The most convenient proportions to arrive at compounds which bear well the action of the rollers, shears, and presses; are not easily oxidable; are sufficiently hard to resist wear; and, above all, have enough intrinsic value, so as not to debase that of the metal made into gold, silver, or copper coins.

For the gold and silver coins, we must employ metals perfectly refined, and alloy them with copper also pure, which imparts to gold and silver, too soft by themselves, the required resistance and hardness.

The standard or fineness of a coin is the proportion of pure metal it contains. The French standard of coins is $\frac{9}{10}$; that of medals is higher, as will be seen:—

For gold coin	90 gold,	10 copper.
" " medals	91.6 "	8.4 "
For silver coin	90 silver,	10 "
" " medals	95 "	5 "

The English standard is about $\frac{11}{12}$. The gold coin contains 11 parts of pure gold and 1 part of copper.

The silver coin contains a greater proportion of pure metal, and is composed of—

Silver	72.5
Copper	7.5
	100.0

Before 1826, silver entered into the composition of the British gold coins. Hence the difference in color of these coins, at various epochs.

The copper coins, manufactured in France since 1852, contain:—

Copper	95
Tin	4
Zinc	1
	100

Previously, their composition had often varied. Nevertheless, zinc was rarely employed; whereas the proportion of tin was sometimes considerable.

The small coins have not only often varied, but their intrinsic value has been singularly changed. At certain epochs, the small coins contained from 1 to 2 parts of silver for 4 of copper. During the revolution, the small coinage was made with all kinds of metals, with scarcely any regard to the standard or quality. Hence, the great variety in the currency which was remelted in 1852.

The old red *sous*, or *sols royaux*, were nearly pure copper. The hard, sonorous, and yellowish-white sous, coined during the Republic with the metal from church-bells, had for an average composition copper 86 and tin 14. The yellow sous, manufactured at the same time with a refined bell metal, were made of copper 96 and tin 4.

The manufacture of coins is at the present time protected by a very efficient system of checks. Skilful chemists are employed at the mint, who, every day, receive samples taken from the beginning, middle, and end of each casting operation, and assay them. The latitude allowed is 0.002, more or less.

It has been proposed to manufacture the new silver fractionary coins of the standard of 835 thousandths. The difference of 65 thousandths in excess of copper, or about 7 per cent. less in the weight of silver, is intended as a compensation for the supposed difference between the nominal and the intrinsic value of these coins.

The alloy of 835 parts of silver and 165 parts of copper is said to be as malleable as the ordinary alloy, but with a somewhat yellower color. Mr. Péligot has proposed to add zinc to this alloy, which would possess all the required qualities with a composition of 835 parts of silver, 93 parts of copper, and 72 parts of zinc. According to Mr. Péligot, such coins are white, elastic, sonorous, and less ready to turn black than the present alloys, on account of the feeble affinity of zinc for sulphur.

The standards of foreign coins are very variable. The silver coins in certain countries, and especially in Germany, are of a very low standard. Some have been made of equal parts of silver and copper. Others, which are more properly called *monnaies de billon* (small currency), contain more copper than silver.

Belgium, the United States, &c., have manufactured coins of nickel, or of alloys of nickel with copper and silver.

The last small fractional coins made in Belgium contain copper 75, nickel 20, and zinc 20.

The small Swiss currency, coined in Paris a few years ago, contained copper, zinc, silver, and nickel. Their nominal value has recently been much lowered.

The new billon coinage of Italy is made of:—

Copper	95
Tin	5
	100

In certain foreign gold coins, gold is brought up to the proper standard by a mixture of equal parts of silver and copper. This alloy expands more than if copper alone were employed, although the specific gravity of gold alloyed with silver differs very little from the average specific gravity of the two metals.

Moreover, we would remark that, as gold, almost always, naturally contains a small percentage of silver, difficult to separate in an economical way, silver is a constituent part of gold coins, which therefore are ternary alloys. It is always possible to bring the gold to the proper standard, but the determination of this standard, on account of the presence of silver, is more difficult than that of silver coins, where copper alone has been added.

At the present time, the bronze for medals is generally made of copper 99 and tin 1. Zinc is rarely added to it. Nevertheless, according to the size of the medals, it is sometimes necessary to change the proportions, which vary between 90 to 95 of copper, and 10 to 5 of tin.

The ancient coins and medals were also based on ternary or quaternary alloys. The numerous analyses, made of coins found in various excavations or collections, have never been concordant, and do not show any constancy or method in the manufacture of the coins.

In certain Roman coins found in Flanders and in the north of France, silver was the predominating metal; in others it was copper. The proportions of tin and gold were comparatively very small.

The coins of antiquity were often manufactured from bronze statues, which the ancients erected and melted again with an equal facility, according to the fickleness of arms and fortune. The gold found in these coins was probably that used for decorating the broken statues; and the tin had quite likely the same origin.

Moreover, the ancients did not know how to refine the compound metals, and their metallurgic knowledge did not enable them to eliminate the foreign elements, which we at present extract by the refining processes.

The old Indian coins, like the Roman ones, were made from quaternary alloys of silver, copper, tin, and gold. In those where silver predominated, the proportion of copper varied from 9½ up to 48 per cent. of the weight of silver.

The Saxon coins were an alloy of copper and tin, with smaller proportions of silver and lead.

Some bronze coins from Attica contain, according to the analysis made at the mint of Paris—

Copper	88
Tin	10
Lead	1.5
Loss	0.5
	100.0

To sum up, the majority of the coins of antiquity, recently analyzed, show the constant and nearly always simultaneous presence of gold, silver, copper, and tin; and that, whether they were gold, silver, or bronze coins. Moreover, a few of these coins have been proved to contain small proportions of lead, iron, or zinc, which metals, the latter especially, were less known or employed, and were only to be found accidentally in the alloys employed in the arts of the earliest ages.

From analyses made at the beginning of this century by the chemist Thomson, the composition of the silver coins of various countries was as follows:—

	Silver.	Copper.
England	92	8
" sterling money	92.5	7.5
Austria	90.5	9.5
Denmark	88	12
Spain	89.5	10.5
"	84.5	15.5
France	90	10
"	1	9
Holland	92	8
Hamburg	50	50
Piedmont	90.5	9.5
Portugal	89	11
Russia	76	24
Switzerland	79	21

III.

ALLOYS FOR PIECES OF ORDNANCE, ARMS, PROJECTILES, ETC.

Pieces of ordnance from the beginning were cast of bronze. The ancient rules prescribed an alloy of 100 parts of copper to 11 of tin.

Numerous experiments have been made, at various times and in different countries, in order to determine exactly the best proportions of copper and tin; yet, notwithstanding these trials, at present we use nearly the same proportions as formerly.

Originally zinc entered into the composition of bronze for cannon, but its use has been gradually discontinued. There was a time when pieces of ordnance were generally made of a mixture of brass and bronze; these two alloys being made separately and then combined.

The brothers Keller have employed the following composition for the pieces of ordnance cast in their foundry:—

Copper	100
Tin	9
Zinc	6
	115

The proportions since admitted among the principal nations of Europe have been:—

	Copper.	Tin.
England	100	12.5
"	90	10
"	88 to 92	12 to 8

And, according to various authors:—

Austria, Bavaria	Copper...100	Tin...10	
Denmark	Copper...100	Tin...10	Zinc...0.125
Spain	Copper...100	Tin...11	
Prussia, Russia, Saxony	Copper...100	Tin...10	

The mining engineers and officers of artillery in France have undertaken many experiments, not only on the binary alloys of copper and tin, but also on the complex alloys of bronze united with iron, lead, zinc, &c. It has generally been found that all these complex alloys were altered by remelting, were difficult to obtain, and required great precautions during the casting, without giving much certainty as to the results.

It has been tried to combine separately first, and then together, the cast iron, copper, and tin, without arriving at truly homogeneous and solid alloys. Therefore, it has been necessary to return to bronze, and to study thoroughly the properties of this metal. We have not to examine here the applications of cast iron, wrought iron, or steel to the manufacture of ordnance.

The alloy of copper and tin, in order to be best suited to the manufacture of ordnance, must present the following characteristics:—

A finely granular fracture, of a reddish tinge, without any admixture of whitish spots.—Yellowish texture.—Specific gravity above the average of the two component metals.—Presenting the maximum of malleability and tenacity possible with the alloys of copper and tin.—Becoming harder and less ductile by hammering.—Receiving, on the other hand, an increase of malleability and ductility by being tempered and dipped into cold water, which is just the opposite of what happens in the case of iron, steel, or cast iron.

One of the most important points for this alloy is as perfect a homogeneousness as can be obtained in practice. Tin renders copper tougher, harder, but more brittle, and has a tendency to separate in the alloy, and to disappear by the effects of heat or friction. If the tin becomes separated from the copper, or is not thoroughly combined, we obtain metallic grains very rich in tin, without adherence, and fusible enough to be liquefied or disintegrated by the action of the powder, leaving the copper in the shape of a spongy mass, without consistence. Thus, it is an acknowledged fact that bronze, remelted several times, becomes more dense, more tenacious, and harder.

It is therefore proper, provided the composition of the alloy be kept constant, to employ old bronzes combined with a certain proportion of new alloy, for the casting of ordnance. In the government foundries, each casting is made, according to this rule, of old bronzes, runners, scraps, and a certain percentage of new metals. For instance, the following proportions are employed:—

22	parts of new copper.
3.3	" " tin.
80.7	" " old broken ordnance.
114	" " runners, scraps, &c.
220.0	

It is useful to know, by previous analysis, the exact composition of each component part; then to determine in what proportions they are respectively to be employed for the alloy. And in order to compensate for the losses which occur during the melting and casting, the proportions are regulated as if the bronze were to contain from 13 to 14 per cent. of tin.

The composition may also be verified during the melting, by means of a rapid analysis effected on a few drops of molten metal, dissolved in nitric acid. The stannic acid is rapidly washed upon the filter, which is then thrown wet into a red-hot platinum crucible. The tin being known, copper is deducted by difference.

The same proportions of copper and tin do not answer for every kind of bore. The proportion of tin must evidently be greater for the large pieces. Thus, we may admit as good proportions: 8 to 9 parts of tin with 100 of copper for No. 8 bore; and 11 to 13 parts of tin with 100 of copper for the bores of 12 and above.

Will the new metals introduced into the arts bring new and unforeseen qualities to the alloys for pieces of ordnance? We cannot answer this at the present time, on account of the incomplete data furnished by experimenters. Is it to be presumed that, with the known metals, the arts will arrive at better alloys than those already obtained? This is probable; but, among the numerous experiments made, nothing allows us to foresee better results than those known at this time.

If progress is not to be sought for in the alloys, we may possibly find it in the processes of combination, such as those which have admitted of the union of aluminium with copper and other metals; which allow of the introduction of a fusible, oxidizable, and volatile metal, zinc for instance, into a more refractory metal, such as copper, which seems to admit only of a certain proportion of zinc by the direct alloy.

Electricity is, without any doubt, one of the powerful agents to which the science of the future will have recourse to obtain, if not an alloy, at least the fixation of certain metals into other. This is a field from which we may expect many important changes and discoveries on the subject of alloys. We confine ourselves to this indication, and pass to the rapid classification of the interesting data furnished by the past and the present, concerning our present study.

The ancients, who were not conversant with the art of working iron, and had scarcely any knowledge of the metal itself, used for their weapons the various alloys of copper and tin known under the name of bronze.

Many of these alloys appear to have been made of 14 parts of tin to 100 of copper. However, it has been found by analysis that certain arms contained from 17 to 18 parts of tin for 83 to 82 of copper.

Roman weapons have shown by analysis—

Copper	81
Tin	19
	100

Other weapons, collected from recent excavations made on the places traversed by the Roman cohorts in ancient Gaul, gave on an average—

Copper	92
Tin	7
Lead	1
	100

Several have exhibited a trace of zinc.

The ancient alloys for weapons or edge-tools appear, most of them, to have been hammer-hardened, after being cast, in order to increase the density and hardness of the metal. The makers of these primitive tools have evidently tried to find in bronze certain of the qualities of steel, which metal was not known to them.

The hardening by a slow and protracted hammering must evidently have imparted to their alloys a greater hardness, and therefore a sharper edge; but the tenacity of the metal would have been impaired, and the weapons rendered brittle, if the ancients had not had recourse to the annealing and dipping processes, which were certainly known, and without which the cold hammered metal would have lost all toughness and suppleness.

We know that if iron and steel become brittle by the hardening process, it is not so with bronze, which, being heated to the proper point and then dipped into cold water, acquires toughness and ductility at the same time.

Arms and cutting instruments have recently been studied by many savans and manufacturers. We have already seen in this work how many metals have been experimentally alloyed with steel in order to improve its cutting edge, to give it a damaskeened pattern, &c.

Gold, silver, platinum, nickel, aluminium itself, and many other metals, have been brought forward to impart to steel peculiar properties. Nothing that we know of at the present time has given sufficiently certain, complete, and secure results to encourage the manufacturers in working them on a large scale.

Therefore we have no such alloys to indicate, and we refer our readers to what we have already said about the possible combinations of steel with the other metals.

We shall terminate this chapter by rapidly mentioning a few alloys adapted to our subject.

The bronze or brass for the mountings of arms, which is said to best fulfil the required conditions of hardness, malleability, and tenacity, is made of:—

Copper	80
Zinc	17
Tin	3
	100

We at present employ, for the same purpose, the alloys of copper and aluminium, and the white alloys in which copper, zinc, and nickel are generally employed.

Alloys for projectiles:—

Lead shot.	Lead	99
	Arsenic	1
		100

In the preparation of lead shot, a little arsenic is added to the lead, which is allowed to fall from a great height, and acquires a more spherical shape, instead of an elongated one. In order to produce the arsenide of lead necessary for the operation, it is sufficient to melt the lead with some arsenious acid. Certain makers employ the ordinary commercial lead, without any preparation; however, from the opinion of the majority of manufacturers, the arsenide of lead is to be preferred.

Gun balls.	Lead	97	98
	Zinc	3	2
		100	100

This alloy is said to give more exactness in firing than is the case with balls of pure lead; but we think that this result requires confirmation. We rather believe that a little zinc added to lead, increases its hardness, and prevents its loss of shape by cooling. Indeed, it often happens that the balls, by the contraction due to the cooling, contain cavities which may be seen by cutting. But zinc, when the alloy is well combined, appears to prevent the shrinkage entirely, or at least partially.

This defect has been obviated by giving an ovoid shape to the ball moulds, and then compressing the cast balls to a spherical form under a press. In England, several manufacturers have tried to obtain the balls from drawn-out cylinders of lead, cut into fragments of convenient size, and then compressed into shape.*

The modifications in the shape of the projectiles, which tend to be substituted for the spherical balls in weapons of war, will bring into use, for the reasons already stated, the alloys of lead and zinc, or zinc alone, especially if the volume of these projectiles be much more considerable than that of the old balls.

IV.

ALLOYS FOR ROLLING AND WIRE DRAWING.

The alloys of the majority of the usual metals, which we have previously examined, may be rendered ductile and malleable by following certain proportions indicated by experience.

In the ordinary practice of the arts, the so-called ductile metals, such as gold, silver, copper, &c.,† when alloyed with other metals, tin, lead, zinc, for instance, may furnish intermediary products, which are ductile and malleable at various degrees.

We shall not examine here all the ductile alloys which may be produced for the rolling and drawing processes. Moreover, the bases of these alloys will be found in various parts of this book.

We shall only speak in general of the preparation of the principal alloys of copper with zinc, tin, or lead,

* Lead is not entirely devoid of elasticity, and this property has prevented the further use of compression in the manufacture of balls. The balls, which, immediately after being compressed, fitted the bore of the gun, had expanded so much after some time of rest in the armories, that they would not enter the same gun.—*Trans.*

† We do not mention iron, which being ductile and malleable when alone, loses these qualities, or at least does not acquire any new ones, when it is combined with other metals.

which are the most useful for the manufacture of plates, sheets, bars, and wires.

A malleable brass was, for a long time, obtained by directly treating calamine in the "German" furnace. It has only been since the beginning of this century that the large foundries have made brass by the direct alloy of copper and zinc in the metallic state.

The distance of the mines of calamine was also the principal drawback to the manufacture of brass in the French works. At the end of the year 1816 experiments were begun at the Romilly works, as we have mentioned in our book *De la Fonderie*, for the direct alloy of copper with zinc, but were not satisfactory for a long time. The metal produced was tenacious enough, but hard and little malleable. Better results were obtained by refining the copper intended for the crucible, because, until then, a portion of the zinc was oxidized and lost in the drosses. But it is to the small proportion of ½ of one per cent. of lead, added to the alloy, that we owe the success of that mode of operation.

From that time, the metal, without losing its tenacity, became milder under the rollers, more ductile in the draw-plate, and wires were obtained as fine as those made from the best brass of Namur.

Mr. Le Brun, at present inspector of the *Ecoles des Arts et Métiers*, was one of the authors of the progress made in the manufacture of malleable brass at the Romilly works, of which he was then the manager. We owe to him the following proportions, which have been the base of all such alloys, without sensible change.

Alloy for hammering, plates, and fine wires:—

Copper	67
Zinc	33
Lead	0.5
	100.5

Alloy for pin wire, which must possess a certain toughness:—

Copper	67
Zinc	33
Lead	0.5
Tin	0.5
	101.0

In general, if we increase the proportion of copper, the alloy is harder and clogs the file more; if the proportion of zinc is increased, the metal becomes less homogeneous and tenacious. The stirring in the crucible must be made with dry white wood, instead of an iron tool, which becomes mixed in the alloy, and renders it flawy and hard.

From these compositions, we see that the brass for rolling, sensibly remains between the limits of 2 parts of copper to 1 of zinc, in the case of brass of first quality. It appears to be demonstrated that a less proportion of zinc will not give a metal as malleable, when hot, as the above alloys, and without the aid of lead and tin.

But it is possible, in the brass of second quality, to employ as much as 40 parts of zinc to 60 of copper. The color of this alloy is a pale yellow, intermediate between that of brass of first quality, and tombac. The fracture of the metal is close and fine; its specific gravity reaches 8.45, whereas by calculation it would give about 8 only, from whence we infer that there is a contraction.

This alloy, which ought to be considered as a chemical compound in definite proportions, is harder than copper, very difficult to break, and so malleable that it may easily be forged when hot, and planed when cold.

We published, a few years ago, a note relative to the process of casting the copper intended for rolling. We shall borrow from it all that we have to say on this subject, nor do we believe that our conclusions should be modified by what has since been done in the works where malleable copper is manufactured.

"The experiments we have made in running cast iron into metallic moulds, have caused us to ascertain whether a similar process would not be advantageous for casting the copper plates intended for rolling. From inquiries made at one of the copper-works of the department of l'Eure, and from our own researches, we have obtained sufficiently satisfactory data on the casting of copper into metallic moulds, to enable us to advise manufacturers to prefer this process, which, in future, will be found more advantageous than those actually employed in the majority of works where sheet-copper is produced.

"The method, which, in the absence of a better one, was preferably employed for casting rolling copper, consisted in pouring the melted copper into moulds of hard stone, covered with an earthy coating, heated upon the stone itself. These moulds, which, moreover, did not produce castings always free from blown holes, and other grave defects, were also exceedingly heavy and difficult to move. Besides, they would become out of shape, on whatever bottom they were resting. The repairs were frequent and costly, on account of the wear due to the shrinkage, notwithstanding the fact that the cast metal was taken off as rapidly as possible.

"The importance of these defects caused a search for better processes, and several manufacturers soon began to employ cast-iron moulds. The melted copper was run first into uncovered moulds resting upon a fixed copper bottom; the whole being heated to a temperature of from 80° to 100° C. This method, which is possibly employed at the present day in some works, replaced with advantage the use of stones, although it is open to the general objection of uncovered and too easily disturbed moulds.

"After numerous and often unsuccessful trials, it became possible to obtain better results with the process which we are going to describe. In our opinion, casting under pressure is the base of the new improvements which are to be sought for.

"The upright standing ingot-moulds, tried with great success in two or three works in the vicinity of Evreux, are made of two cast-iron pieces, perfectly planed, and inclosing a space equal to the metallic slabs desired, but not less than 0.012 metre in thickness. On the top is an opening, like a funnel, for running in the metal, and for the escape of gases.

"The side of the funnel opposite that for the entrance of the copper is somewhat higher, in order that the liquid shall not run over. Each mould is kept closed by clamps or wedges, and is inclined during the casting about ten degrees.

"The moulds are subjected to the following necessary operation before casting: they are smeared over with just enough oil to retain a very thin layer of charcoal-dust, which is thrown upon it by means of a sack similar to that used by moulders in sand. The temperature of the moulds also requires attention, as more than from 80° to 100° C. will impair the homogeneousness of the alloy; a less heat will occasion flaws, blown holes, and separated drops. The workman in charge of the moulds must be careful to open them immediately after the casting is done, otherwise the slabs will be broken. The same person attends to the cooling of the moulds, when, after each operation, they have acquired too high a temperature.

"As regards the cast-iron moulds, experience has taught that the metal must be very mild, and in every case well annealed. The moulds which have not been annealed, generally produce copper plates or slabs filled with blown holes.

"But although these processes are to be preferred to the old methods, they may yet be considerably improved. For instance, while we retain the principle of casting under pressure in metallic moulds, we may vary the nature of these moulds, and obtain more homogeneous and perfect metallic slabs or plates which are better fitted for the purpose of rolling.

"Metallic moulds made of brass (copper 70, zinc 30), oiled and then smoked with rosin soot, have furnished plates without blown holes, but presenting a few blemishes at the upper part. The moulds become heated very much and crack.

"Cast-iron moulds, perforated with small holes for the escape of the air, at the same time that they retained the clay with which the inside was covered, gave us better results. The clay used was the fine stuff employed by moulders in clay, and its thickness was not over two to three millimetres, regulated by a board. This clay was then brought to a red heat, and covered afterwards with a coat of the liquid black employed by cast iron moulders. The copper plates obtained from such moulds were very fine and without any blown holes. It remains to be ascertained whether the pellicle which covers the metal, and which is thicker that that of the metal cast in direct contact with the metallic moulds, will not prevent the thorough scouring necessary for a fine appearance in the laminated sheets. Once this fact is ascertained—and we have no doubt that it will succeed with the alloys of copper and zinc*—the process which we have

* The results will not be so advantageous for pure copper. This metal, employed in the pure state, and cast in sand, loses part of its tenacity, and becomes very flexible and porous, especially if the castings are not very thick. It may be feared that the lining of clay, notwithstanding its thinness, will act the same as sand on the quality of copper.

indicated will be the best, because all of the inconveniences resulting from the direct contact of the metallic surfaces will be avoided without considerably increasing the expenses of labor and repair. Copper moulds with a lining of sheet-iron, or cast-iron moulds alloyed with 5 per cent. of copper, well annealed and maintained at a proper temperature, gave also good copper slabs for rolling; but none of these latter moulds have, as completely as those lined with clay, prevented the formation of blown holes.

"This, the most troublesome of defects, particularly so for rolling copper, is corrected in the preparation of the alloys.

"Pure and new copper is naturally porous during the first meltings, but becomes improved by repeated fusions. Nevertheless, it is very difficult to obtain sound slabs of pure copper, and it has been found advantageous in practice to add from 1 to 2 per cent. of lead to the copper which is to be laminated. A small percentage of lead is also very proper for brass, and excellent sheets are made of 66 parts of pure copper, 33 of zinc, and 1 of lead. The manufacturers of these alloys sometimes carry their economy so far, by reducing the proportion of copper, that the proper products cannot be obtained. There are limits within which it is prudent to remain, and the proportion of copper should never be less than 60 per cent. The alloys of brass, of similor, &c., like new copper, become improved by a second fusion; but when the direct alloy is properly made, that is, when the metals are combined after having been separately melted, and when the proper degree of heat is obtained, the stirring sufficient, and the casting rapid, good products may be obtained without incurring the expense and waste of a second melting.

"Old pieces of copper added to the new alloy help the combination of the metals; but the old pieces must be of good quality, and generally sheets deprived of any trace of solder, tin, or iron. Old kitchen caldrons, saucepans, pipes, &c., are not good, because they are seldom pure; when employed, they are previously submitted to a red heat, in order to eliminate most of the foreign metals or substances. The old copper sheathings of ships are not satisfactory; the cast plates made of them are exceedingly hard and brittle, and experience has proven that these slabs or plates remain of an inferior quality, even after the addition of 50 per cent. of new copper. It is therefore necessary to make a good choice of the old copper which is to be added to the alloy, since it has a great effect on the results. The best old copper comes from stamped, drawn-out, and laminated pieces, from the waste of laminated sheets or imperfect plates, and with them we obtain more homogeneous and tenacious alloys, which, therefore, are better adapted to the laminating process.

"We must carefully verify, before they are introduced into the alloy, the old pieces of copper cast in foundries, because these coppers have a variable composition, and are almost always the result of all sorts of old copper thrown into the crucible, without regard to their quality. Indeed, the ordinary castings do not require an alloy as rigorously exact as is the case when the metal is to be laminated.

"To sum up, the manufacture of the copper, brass, and other similar alloys for rolling, is based upon:—

"1. The process of casting, the material, shape, and size of the metallic moulds, which receive the molten metal, and we have stated the conditions which, in our opinion, are to be attended to.

"2. The quality of the raw materials and the composition of these alloys. This question is most important, and it is necessary to determine in advance what will be the most favorable and economical conditions for the mixture of new copper with zinc, tin, lead, or old copper.

"3. The mode of operation, and the proper degree of temperature for casting. Copper and its alloys require generally to be cast hot, nearly in a state of ebullition, if we desire to obtain sound castings; nevertheless, we should not go beyond certain limits if we wish to avoid waste. The proper time for casting is, as a rule, when the surface of the bath becomes bright, *slides* to a reddish-white, and shows by its motion that the molten mass has acquired the maximum of temperature which is convenient."

Among the alloys used in the arts for rolling and drawing, we would indicate the following compositions, which we shall examine again further on:—

Bronze for sheathing—

Copper	96
Tin	3
Zinc	1
	100

Brass plates, called Jemmapes brass—

Copper	64.6
Zinc	33.7
Lead	1.5
Tin	0.2
	100.0

Similor for gilding or plating—

Copper	92.7
Zinc	4.6
Tin	2.7
	100.0

Maillechort for rolling—

Copper	60
Zinc	20
Nickel	20
	100

V.

COPPER ALLOYS FOR SHIP SHEATHINGS.

Mr. Bobierre, Professor of Chemistry at Nantes, has paid a great deal of attention to the causes of alteration in the bronzes employed for sheathing ships, and to the process for obtaining these bronzes in the best possible conditions of alloy and manufacture.

We here sum up rapidly the observations of Mr. Bobierre, which will be found sufficient to elucidate the question of these sorts of bronzes.

Pure copper and zinc are yet employed for sheathing ships; but the experiments of Mr. Bobierre, made on samples of sheathing which had been exposed to the action of the sea for several years, have brought him to the conclusion that bronze is preferable as regards solidity and duration.

As a rule, it is desirable that the sheathing bronzes should be made of copper and tin, with a minimum of 4 per cent. of the latter metal. The best proportions appear to be 5 or 6 per cent. of tin.

According to Mr. Bobierre, we may consider the molecules of such homogeneous alloys as so many voltaic couples, from which the sea-water has a tendency to eliminate tin, in preference to copper. On the other hand, the force of cohesion being greater in bronze than in pure copper, the alloy ought to resist better the action of sea-water.

The noted results of trials made in France and in England, on the sheathing of vessels which had made long voyages, show that good bronze alloys had resisted in the proportion of 2 to 1, and 3 to 2, as compared with sheathings of pure copper, or of copper alloyed with from 1 to 2 per cent. of tin.

The alloys of a very red color, that is to say, which do not contain enough tin, are heterogeneous, scorified, and with a coarse and irregular grain. This is explained by the difficulty of thoroughly combining a very small proportion of tin with a large mass of copper, notwithstanding a good fire and complete stirring. Therefore, in such alloys too small a proportion of tin causes blown holes and stains, where it ought to act as the electro-positive element in opposition to copper.

Mr. Bobierre has found by analysis that the sheathing bronzes contained not only sensible traces of arsenic, but also a comparatively large proportion of lead. These facts will be explained—first, by the ordinary presence of arsenical iron, and arsenic itself, in the tin oxides of Cornwall and of the coasts of Brittany; second, by the necessity of aiding the difficult rolling of pure alloys of copper and tin, by an addition of a few hundredths of lead.

The bronze sheathing of the ship *Sarah*, which had imperfectly resisted the action of sea-water, was found by Mr. Bobierre to contain—

Copper	950 to 970 parts.
Tin	25 " 35 "
Lead	5 " 13 "
Arsenic	perceptible traces.

On the other hand, those of the packet-ship *Ferdinand*, which had stood very well, were composed of—

Copper	850 to 950
Tin	41 " 45
Lead	6 " 9
Arsenic	traces.

Samples from the ship *Aline*, which had made several long trips, without any alteration of her sheathing, gave—

Copper	935
Tin	55
Lead	10
Arsenic	trace.
	1000

Other samples, taken by several manufacturers and ship-owners of Nantes from well-preserved sheathings, gave a proportion of tin varying from 55 to 65 parts per thousand parts of alloy.

Mr. Bobierre concludes, from these facts:—

That tin, which plays the part of an electropositive metal, enters in too small a proportion into the imperfect alloys;

That, up to a certain point, it is possible to determine a ratio between the proportion of the more oxidable metals and the propensity of the alloy to become altered;

That the sheathings which had shown a great power of duration contain at least 4 per cent. of tin;

Lastly, that the grain of the alloy is coarse, its color bad, and the stains of tin apparent; or, to sum up, that the tin is not uniformly divided through the mass, when its proportion is below 4 per cent.

These facts being admitted, and if we remember that when an alloy of copper and tin is melted, the latter metal is oxidized in preference to the former, we may then admit that the experiments of Mr. Bobierre, without having a rigorous exactness, which is not, however, claimed by this chemist, may serve, *a priori*, as the basis for the production of good bronze sheathings, which a ship owner has the right to expect.

Experiments made on bronze sheathings, allowed to stand for a certain length of time in a solution of—

Alum	40 parts
Cream tartar	20 "
Common salt	40 "

have shown to Mr. Bobierre, besides his analytical results: that sheathings rich in tin, with a color similar to that of bronze ordnance, a fine grain, and a fine homogeneous appearance, had their thickness uniformly diminished;

That the bronzes deficient in tin, and with the appearance of bad bronze, were unequally corroded, sometimes rough to the touch and sometimes perforated, but most generally presented large worn surfaces, and irregular-shaped stains.

Trials made on a larger scale, have confirmed the laboratory experiments of Mr. Bobierre, and we may conclude:—

That bronze sheathings, as regards stability and duration, are to be preferred to copper and brass sheathings;

That the irregular alterations, so ruinous to shipowners, result from a defect in the manufacture of these bronzes;

That the presence of arsenic in these bronzes does not produce as rapid an alteration as is the case with pure copper;

That the sheathing bronzes, with only from 2 to 3 per cent. of tin, are not homogeneous, and are irregularly altered; and that their durability on the ocean is, in every case, much inferior to that of the bronzes holding from 4.5 to 5.5 per cent. of tin.

The desire to do the rolling economically by diminishing the hardness of the alloy, and the introduction of harsh copper, of a doubtful quality, are the causes of the inferiority of the low standard bronzes employed for trading ships.

The addition of a small proportion of lead, and even of zinc, into bronze sheathing, will improve these alloys by aiding the thorough distribution of the electro-positive element in the metallic mass.

If, during the service at sea, the bronzes are a little more subject to fouling than pure copper, according to certain captains, the inconvenience is not so great, with good alloys, as to prevent the employment of bronze sheathing.

As for the use of pure zinc sheathing, all navigators know with what rapidity and energy the parasite molluscs (barnacles, &c.) stick to that metal, and render its employment impossible.

Whatever be the duration and the cheapness of zinc, it will always be more advantageous to prefer bronze, brass, or even pure copper, notwithstanding the constant alterations due to the frequent impurity and to the thinness of the latter metal.

A few chemists have recommended the alloys of tin and zinc, in substitution for pure zinc sheathing. These alloys are hard, difficult to roll, and do not appear to give better results than pure zinc; moreover, their greater cost counterbalances their possible advantages.

Among the other alloys proposed or employed for sheathings, we may notice the *alloy of Muntz,* which is made of—

Copper	56 parts.
Zinc	40.75 "
Lead	4.50 "
	101.25

According to Mr. Muntz, the lead plays an important part in this alloy, which without it would not be sufficiently oxidizable to prevent the careen from fouling. This alloy may contain more or less copper, and therefore be more or less economical. At all events, the proportion of copper should never be less than 50 per cent.

This alloy appears to have given satisfactory results; but Mr. Bobierre does not think it so good as a bronze made under good conditions. In this respect, this chemist disagrees with many ship-owners, who prefer the copper-zinc sheathings to every other alloy, even those of copper and tin made according to the indicated rules.

VI.

ALLOYS FOR TYPE, ENGRAVING PLATES, ETC.

According to Mr. Ch. Laboulaye, an authority in such matters, an alloy for type metal must fulfil the following conditions:—

1. Not too great a propensity to crystallization, otherwise the metal will crystallize near the metallic surfaces of the mould;

2. Ready fusibility, in order to keep the metallic bath at a proper temperature without too much oxidation, which may be rapidly produced by the frequent dippings of the casting-ladle;

3. Sufficient hardness for preventing the crushing of the letter, while printing; and at the same time sufficient softness for facilitating the operations following the casting, and the printing itself;

4. A reasonable cost, so as not to increase beyond measure the value of printing material.

It results from these conditions, that lead has been considered, up to the present time, as the base of alloys for types. However, as it requires to be hardened, its combinations with brittle metals have been tried.

Zinc has the advantage of cheapness and easy fusibility; but at the low temperature necessary to insure its combination with lead, it remains pasty and does not fill the moulds.

The preference has therefore been given to antimony, which, alloyed with lead, answers the purpose better.

The alloys of lead and antimony, which contain from 10 to 30 per cent. of the latter metal, according to the degree of density desired, may be made as brittle as desired by increasing the proportion of antimony. As long as the proportion of antimony is not over 15 per cent., these alloys possess a property of expansion which is very advantageous for sharp casts.

The alloy with 15 per cent. of antimony is the most satisfactory, as regards fluidity and expansion by cooling. It is more fusible than either of the component metals.

However, it was ascertained that the alloy of lead and antimony, notwithstanding its proper degree of hardness, had a vitreous nature, and imperfectly resisted the action of the press and of the scouring caustics; it was then tried to increase the resistance without losing the other qualities of the alloy. This result was obtained by the employment of tin or bismuth.

The proportion of tin appears to range from 6 to 8 per cent. A greater amount would cause a waste by oxidation; and the alloy would be brittle, by the too great tendency of tin and antimony to crystallize.

Various alloys of copper and zinc have been tried, but without satisfactory results.

MM. Didot have employed for their stereotypes an alloy of 1 part of copper, 9 of tin, and 100 of the alloy of lead and bismuth. Mr. Laboulaye has used for the same purpose an alloy of 1 part of copper, 6 of tin, and 100 of type-metal. But these alloys have not been successful, on account of their high price, their hardness when needing repairs, their refractory character and rapid oxidation, and, lastly, their tendency to crystallization.

Mr. Laboulaye indicates an alloy of tin with from 1 to 2 per cent. of iron, which, being added to the type-metal in the place of 1 part of lead, gives a compound not very crystallizable, quite hard, and resisting well hard work, such as the printing of newspapers. The same author also mentions an alloy of Mr. Colson, made of equal parts of tin and zinc, which was very satisfactory as to resistance, but was discarded on account of the destruction by the zinc of the iron moulds and matrix, and the difficulty of dressing the types with the knife.

The following combinations are given, more as guides for the experimenter than as absolute bases:—

Printing-types—

Lead	4 parts.
Antimony	1 part.
	5

Small types and stereotypes—

Lead	9 parts.
Antimony	2 "
Bismuth	2 "
	13

Or—

Lead	16 parts.
Antimony	4 "
Tin	5 "
	25

Plates for engraving music—

Tin	5 to	7.5
Antimony	5 to	2.5
	10	10.0

This alloy is the more brittle, as the proportion of antimony is greater. Its specific gravity is less than that of each of the component metals.

Lead	16
Antimony	1

The presence of antimony is sufficient to impart to this alloy a great tenacity. The specific gravity is above the average of the two metals.

This last alloy has been tried in all proportions, from 4 to 16 parts of lead to 1 of antimony. Sometimes tin, zinc, or copper has been added to it; and among several compositions, we may indicate the following ones:—

Lead	8
Antimony	2
Tin	1.5
	11.5 parts.

Lead	4
Antimony	2
Zinc	1
	7 parts.

Lead	7.5
Antimony	2.5
Copper	0.5
	10.5 parts.

For large type, ectypes, matrix, &c., the following proportions have been tried:—

Lead	10
Copper	2.5
	12.5 parts.

Lead	9
Antimony	1
Arsenic	0.5
	10.5 parts.

Copper	8
Tin	2
Bismuth	0.5
	10.5 parts.

Copper	2
Tin	2
Bismuth	2
	6 parts.

Copper	73
Zinc	27
	100 parts.

Copper	5
Zinc	67
Tin	25
Nickel	3
	100 parts.

Tin	12
Zinc	16
Lead	64
Antimony	8
	100 parts.

Tin	56	37.5
Lead	42	60
Antimony	2	2.5
	100 parts.	100.0 parts.

The last two alloys have been employed for engraving plates.

VII.

ALLOYS FOR BELLS, MUSICAL INSTRUMENTS, ETC.

The alloy for bells, known under the name of bell-metal, is generally composed of—

Copper	78
Tin	22
	100 parts.

This alloy is of a yellowish-white color, hard, brittle, difficult to file, and with a crystallization without lustre. It acquires a certain malleability when it is rapidly cooled off, whether by immediate exposure of the casting to the air, or by being dipped into water.

From analyses of old bells made by modern chemists it has been found that the proportion of tin varied from 20 to 26 parts to 100 of copper.

These bells were rarely manufactured with new or pure metals; therefore, the analyses have often shown the presence of foreign compounds, useless or detrimental to their qualities, especially certain white metals, such as zinc and lead. The former metal, when in small proportion, may not really prove a defect in bell-metal. It has even been tried purposely in certain alloys. Indeed, although zinc neither improves the quality nor the sonorousness of the alloy, it does not act very badly, and allows of the manufacture of cheaper bells, which, however, are not so perfect as those made of copper and tin alone. It is not so with lead, which, if present even in a very small proportion in bell-metal, will impair its sonorousness and hardness. Therefore, lead must be avoided at all events.

We do not see any serious objection to the introduction of zinc into the bell-metal, provided that too much of it be not added. A small proportion of zinc renders the alloy more homogeneous, dense, fluid, and ready to acquire the peculiar tint of old bronze.

It also gives a more economical metal, which explains the sensible reduction in the price of bells, at present manufactured on a large scale in certain works. These manufacturers will soon crush the strolling melters, who for centuries had the monopoly of the casting of bells.

The new manufacturer of bells tries to work rationally, analyzes and experiments with various compositions, in order to apply the metals to the best advantage. In the past, on the contrary, there were no other rules than that of the thumb; and old metals were employed, such as broken kitchen utensils, spigots, tinned copper with solder, &c., which could give but dubious results.

If we add to that the want of precise data as to the proportions, the alteration by fusion of the alloys of copper and tin, &c., we must not wonder at the differences shown by the analyses of various bells. These variations were ascertained, especially during the crisis of the French revolution, when the church-bells were taken for the manufacture of cannon and coins.

Besides copper and tin, the presence of zinc and iron was often detected, and also, but not often, that of silver and gold. The presence of the latter metal was less frequent than is generally supposed.

If some credulous minds, at certain epochs, have brought precious objects of gold and silver to be added to the bell-metal, in order to gain indulgences or to make a pious offering, we must believe that the founders were smart enough to pass the valuable offerings through a less ardent fire than that of their furnaces.

As witness, the celebrated bell of the belfry of Rouen, known under the name of the silver bell, and which was believed by tradition to contain an enormous amount of silver. Its analysis, made by the learned chemists of the Paris mint, gave:—

Copper	71
Tin	26
Zinc	1.8
Iron	1.2
	100.0

and not a trace of silver.

As we have already said, it is difficult to preserve the ultimate proportions of bell-metal, which is also true of all alloys. It is therefore necessary to increase the proportion of tin, if we desire that the alloy should have the composition demanded. But, whatever be the excess of tin added, we can never arrive at a perfectly exact composition, on account of the oxidation during the fusion, variable with the fire and the shape of the furnace, and of the phenomenon of separation, which takes place in the mould if the metal has not been well stirred and properly cast.

From experiments on samples of bell-metal, made at different times, we have ascertained variations in the alloy, ranging from 18 to 35 parts of tin for 100 of copper.

In order to counterbalance the loss of tin in the alloy, we believe that without increasing the proportion of tin, a bell-metal might be composed of—

Copper	79
Tin	23
Zinc	6
	108 parts.

If we suppose that the fire is properly managed, and that no unforeseen accidents take place during the melting and the casting, the cast bells ought to have an ultimate composition of—

Copper	78
Tin	20
Zinc	2
	100 parts,

which corresponds to a hard, tough, and slightly malleable metal, the sonorousness of which has not been sensibly changed by the presence of the zinc.

The quality of bells, in regard to sound, resistance, &c., also depends upon the shape and the particular processes of moulding and casting, outside of the question of the alloy. On this subject we refer our readers to our book *de la fonderie* (on foundries).

Zinc, and even lead, are employed in England for the casting of bells; but if the latter metal is tolerated at all, the proportion must be exceedingly small, just enough to perfect the homogeneousness of the alloy.

Several analyses of modern English bells give, on an average—

Copper	80
Tin	11
Zinc	6
Lead	3
	100

In old bells of the same country, an exaggeration of tin has been found, as much as 40 per cent. of the alloy. These bells were exceedingly thick, and their shape was widely different from the forms recognized by our present founders.

In France also the proportion of the white metals, such as tin and zinc, is exaggerated, especially in the alloys for hand-bells, clock-bells, &c. For such objects the common alloy employed is a sort of *potin* (yellow pewter) made of—

Copper	55 to 60
Tin	30 to 40
Zinc	10 to 15

The metal for gongs and cymbals is composed, on an average, of—

Copper	75
Tin	25
	100

This metal is whiter, more sonorous, more brittle than bell-metal, and is not so easily filed.

Chinese gongs, analyzed by Mr. Darcet in 1832, have shown 78 parts of copper to 22 of tin, and a specific gravity =8.815.

The composition for cymbals, admitted in the shops of the School of Châlons, after the experiments by Mr. Darcet, was—

Copper	80.5
Tin	19.5
	100.0

These alloys are brittle, and cannot acquire the desired resistance and sonorousness, unless they are dipped into cold water after being heated up to a certain point.

The alloys of copper and tin possess the property, which we have already mentioned, of becoming very malleable after having been brought up to a red heat and immersed in cold water. This property is made use of in the manufacture of gongs and cymbals.

These instruments, cast in a slightly wet and loose green sand, in order to avoid any fracture by shrinkage, are then brought up to a red heat and dipped into water with certain precautions. After this operation they may be forged and hammered. The proper pitch is imparted to them either by the tempering process, or by a more or less protracted hammering at certain places, or by annealing them after they have been hardened by the hammer.

The honor of the discovery of the processes which have permitted of the manufacture in France of gongs and cymbals, has been awarded to Mr. Darcet. The labors of this gentleman are already considerable enough, to make it unnecessary to attribute to him the industrial improvements due to the experience of workers not so well known. Mr. Darcet has certainly made analyses of the alloys of gongs and cymbals, and has given some sound advice; but the processes of manufacture and their improvement are due to the researches of founders, and among them, of Mr. Maillard, the skilful, learned, and modest manager of the foundry shop of the School of Châlons, who has made many improvements in founding and in alloys, and has paid special attention to the processes for moulding, casting, tempering, and hammering the alloys which we have mentioned.

VIII.

ALLOYS FOR PHILOSOPHICAL AND OPTICAL INSTRUMENTS.

(Especially Speculum Metals.)

Without speaking of the white metals, of the *maillechort* (German silver), aluminium, platinum, &c., which are in daily use for the manufacture of certain philosophical or optical instruments, we shall here confine ourselves to the summing up of the best known alloys corresponding to the title of this chapter.

The greater number of these alloys are for the fabrication of metallic mirrors, in which we require a true white color, a fine lustre when polished, hardness, and a clean surface which becomes with difficulty scratched, altered, or tarnished.

The Chinese mirrors, which have attracted the attention of savans, in order to learn the bases for such compounds, have been found to contain sometimes copper, lead, and antimony; sometimes copper, tin, and lead. The latter alloy is grayish, susceptible of a fine polish, but presents no peculiar qualities. Its composition is generally—

Copper	62
Tin	32
Lead	6
	100

The former has a whiter color, a finer polish, and is not so easily tarnished by contact with the air. Its average composition is—

Copper	80
Lead	10
Antimony	10
	100

Certain mirrors of antiquity show—

Copper	62
Tin	32
Lead	6
	100

In France similar mirrors have a composition ranging between—

Copper	66	63
Tin	33	27

These compounds are very hard, brittle, with a fine polish of a steel-white color, and with a lamellar, gray, and dull fracture.

Other more complex alloys have been employed, such as—

Copper	10
Tin	10
Antimony	10
Lead	50
	80 parts.

Copper	32
Tin	50
Silver	1
Arsenic	1
	84 parts.

In addition to these alloys, which are made of ordinary metals, but do not answer all the desired conditions, let us mention a few combinations made by chemists with less known metals, or metals difficult to alloy.

The alloy tried by Mr. Despretz for mirrors is—

Steel	90
Nickel	10
	100

This alloy is very hard, scarcely alterable by the air, and has a specific gravity = 7.684. The difficulties attending its manufacture prevent its application to the arts.

The same chemist has also indicated for the same uses the alloys of palladium with gold or silver. An alloy of—

Palladium	50
Silver	50
	100

has a grayish shade, and is harder and less fusible than

silver. Its polished surface is whiter than platinum, and its specific gravity is about 11.29.

It is said that this alloy, recommended for the manufacture of marine instruments and the scales of thermometers, has been employed for the great graduated circle of the Observatory of Paris.

However, this point is not perfectly settled, some authors contend that the same circle is made of equal parts of palladium and platinum; others, that the alloy is one of palladium and gold, a small proportion of palladium being sufficient to impart a white color to gold, and to increase its hardness.

At all events, it appears to be certain that palladium is a component part of the alloy, and has imparted, whether to gold, silver, or platinum, a certain whiteness and hardness at the same time.

Various chemists, and among them MM. Stodart, Faraday, and Dumas, recommend for the manufacture of mirrors for telescopes (speculum metal), or of objects requiring a perfectly neat polish, the following compounds—

Platinum	60
Copper	40
	100

which has the same color as platinum, and acquires a very brilliant polish.

Platinum	50
Steel	50
	100

which has a remarkable polish, difficult to tarnish.—Specific gravity, 9.862.

Platinum	50
Iron	50
	100

which is crystallized, very hard, and sufficiently fusible. It acquires a fine polish, and does not tarnish.—Specific gravity, 9.862.

Platinum	10
Steel	90
	100

whiter and harder than platinum.—A better polish.—Specific gravity, 8.10.

Platinum	20
Copper	80
Arsenic	0.5 to 1
	100.5 to 101

ought to give the best mirrors, the alloy being more easily effected.—The alloy is of a grayish-white color, acquires a fine polish, does not tarnish, but its lustre is not equal to that of the entirely white metals.

Platinum	60
Iron	30
Gold	10
	100

which is white and does not tarnish, when polished.

Gold	50
Zinc	50
	100

which is whitish, finely granular, and oxidized with difficulty.

Steel	50
Rhodium	50
	100

which is very well adapted for mirrors, according to MM. Stodart and Faraday.—A very fine polish, which does not tarnish.

Platinum	10
Iridium	90
	100

which, according to Mr. Gaudin, possesses more brilliancy than pure platinum.—Not oxidizable.—Becomes harder by the usual hardening process.—May be obtained in sheets for the plating of reflectors.

The alloys of platinum and iridium are very refractory, and may be employed, according to the same author, for the manufacture of crucibles and retorts for chemical analyses effected at a very high temperature.

Tin	29
Lead	19
	48

This alloy, when melted, will adhere to the polished surfaces with which it is in contact, and leave them on cooling. The thickness of the deposit is regulated at will by the time of contact. It is used for making metallic mirrors, and other pieces with facets, which project a dark lustre, and are known under the name of *Fahlun brilliants*.

We certainly pass over many, and possibly valuable, alloys; but the indications which we have just given will show in what direction experimenters have worked up to the present time, in order to arrive at such metallic combinations as will take the best polish, conjointly with the lustre, whiteness, and hardness required for philosophical and optical instruments.

IX.

ALLOYS FOR JEWELRY, GOLD AND SILVER WARES, BRITANNIA WARE, ETC.

The jewelry trade combines the gold ingots, which have a fineness of about 1000 thousandths (24 carats), with various alloys, in order to arrive at the legal standards, and also at the various colors of gold required by the trade.

The three legal standards for jewelry gold, as prescribed by law, are in France:—

I. *First standard* or *high standard gold.*—920 thousandths, or 22$\frac{1}{32}$ to $\frac{1}{2}$ carats, the unit being divided into 24 carats. This standard is more particularly employed by the goldsmiths.

II. *Second standard* or *standard gold.*—840 thousandths, or 20$\frac{5}{32}$ and $\frac{1}{2}$ carats.

III. *Third standard* or *common gold.*—750 thousandths, or 18 carats.

The tolerance is 3 thousandths, one way or the other.

For the inferior standards, or *low gold,* the fineness varies from 500 to 750 thousandths.

The colors of the gold used in jewelry work are:—

Yellow or *antique gold.*—Pure gold.

Red gold.—Pure gold 750, copper 250.

Green gold.—Pure gold 750, silver 250.

Gold feuille morte (dead leaf).—Pure gold 700, silver 300.

Gold vert d'eau (water-green).—Pure gold 600, silver 400.

White gold, sometimes *electrum.*—Gold whitened by a greater or less proportion of silver.

Blue gold.—Pure gold 750, iron 250. This alloy is quite difficult to produce, and is prepared with iron-wire dipped into the molten gold. It is then cast, hammered, in order to make it tough, and afterwards laminated or passed through the draw-plate.

The alloys of gold must be very homogeneous; therefore they are melted several times. A good alloy should not show any cracks or grains when it is hammered or laminated. If the alloy is brittle or harsh, it is rendered softer or milder by melting it with a certain quantity of flux (borax or saltpetre).

The silversmiths employ silver at two legal standards (in France):—

The first standard is 950 thousandths, and the second, 800 thousandths.

The tolerance is 5 thousandths.

The silver employed for the alloys is pure silver, and the standards are well kept.

Thanks to the legal standards required by the French government for the works of gold and silver, and thanks also to the obligatory assays previous to the stamping of these metals, the jewelry, gold, and silversmith's wares manufactured in France offer a better guarantee of quality than similar articles manufactured in England, Germany, &c. In these countries the precious metals, not being subjected to any control, are the object of the most audacious swindles, so much so, that articles sold as gold or silver, will often contain scarcely a trace of these metals.

A quantity of jewelry has been, and is yet, manufactured in England, from gold at the standard of 12 carats and less, alloyed with zinc, instead of silver. This gold, which has nearly the color of 2-carat gold, has no other use than to deceive the trade and the public. Chains, thimbles, pencil-cases, &c., have often been made of this fraudulent alloy, which, after a certain use, becomes separated as though under galvanic action, and leaves the articles entirely useless.

The alloys employed in England for imitating or falsifying gold are generally kept within the limits of the following alloys:—

Jewelry Gold.

Pure gold	38.85
Silver	5.70
Pure Copper	10.20
	54.75

Ring Gold.

Gold (coin standard)	49.60
Pure silver	12.30
Refined copper	23.60
	85.50

Gold (value 45 to 50 francs for 28 grammes).

Gold (coin standard)	31
Pure silver	38
Refined copper	27.5
	96.5

Common Jewelry.

Refined copper	3
Old Bristol bronze	1
	4

plus 25 parts of tin for 100 parts of copper.

If this alloy is to receive a fine polish, the tin is replaced by a compound of lead and antimony. By increasing the proportion of this compound, or diminishing that of copper, the color of the alloy will become proportionally whiter.

Yellow Metal for Dipping.

Copper	7	Bronze* (Copper 7, Tin 2, Zinc 3)	2
Tin	2		
Zinc	3		
		Copper	1
			3

plus 10 parts of tin for each 640 parts of copper.

* We generally call *bronzes* the alloys of copper with tin, even with the addition of zinc and lead. On the other hand, *brasses* are the alloys of copper with zinc, or with zinc and lead, but without tin.

Another Metal for Dipping.

Copper	48
Zinc	15
	63

When in the preceding alloys we employ antimony instead of zinc or tin, the proportion of the former metal ought to be very small, otherwise the compound will be very brittle.

Metal for Gilding.

	Copper	4
Copper 3, Zinc 1	Brass	1
		5

plus 70 parts of tin for each 80 parts of copper.

Manheim Gold.

	Copper	10
Copper 3, Zinc 1	Brass	1.4
	Tin	0.1
		11.5

Or:—

Copper	3
Zinc	1
Tin	0.5
	4.5

Chrysocale.

Copper	9
Zinc	8
Lead	2
	19

Tombac or *Similor.*

Copper	8.
Tin	0.5
Zinc	0.5
	9.0

Red Similor.

Copper	5.5
Zinc	0.5
	6.0

White Similor.

Copper	6.50 to 7
Arsenic	0.25 to 0.5

The two metals are put together in the crucible, and melted while the surface of the bath is covered with common salt in order to prevent oxidation.

For a *whitened copper* we may also employ:—

Copper	24
A Neutral Salt of Arsenic	1.5
	25.5

melted together with a flux of calcined borax, charcoal-dust, and powdered glass.

Bath Metal.

Copper	3	Brass	48
Zinc	1		
		Zinc	13.5
			61.5

Or another:—

Copper	75
Zinc	25
	100

Pinchbeck or *Prince Robert's Metal.*

	I.	II.
Copper	90	30
Zinc	30	60

The two proportions bear the same name; however, the alloy II. is the one most usually known in England under the name of Prince Robert's metal.

The English manufacturers, especially those of Sheffield and Birmingham, employ a great number of alloys, either for counterfeit jewelry, or for many articles of legitimate trade, such as buckles, window fixtures, pieces of hardware, locks, &c., in which they excel, not only by the finish or the good taste, but by the metallic appearance of these wares. We shall also indicate the following compounds, which may be useful to know, whether as metals imitating gold, and for gilding, or as metals imitating silver, and for silvering.

These alloys are well known in France, but not so generally as in England and Germany.

Argentan (*packfund* or *packfong*) *of Sheffield.*—This ordinary quality has a yellowish tinge, and is employed for wires and common articles:—

Copper	8
Nickel	2
Zinc	3
	13

A superior quality, known as *white packfong*, imitates the silver of 750 thousandths, and is employed for spoons, forks, ornamental table pieces, &c.:—

Copper	8
Nickel	3
Zinc	3.5
	14.5

The following alloys are very malleable, white, and susceptible of a fine polish:—

	I.	II.	III.
Copper	4	2	1
Nickel	1	1	1
Zinc	1		

These compounds resemble the alloys made in France under the name of maillechort. Their white color renders them well adapted for the operation of silvering, and there is so slight a difference between their color and that of silver that the body metal is not apparent after scratching or chiselling.

German silver is made of—

Copper	2
Nickel	1
Zinc	1
	4

Chinese white copper or *Chinese packfong*:—

Copper	10.4
Nickel	31.6
Iron	2.6
	44.6

German silver for rolling:—

Copper	6
Nickel	2.5
Zinc	2
Lead	0.3
	10.8

The French manufacturers employ for false jewelry the *Ruolz alloys*, the compositions of which vary between—

Silver	20 to 30
Nickel	25 to 30
Copper	35 to 50

These proportions are those adopted by Mr. de Ruolz; but, by varying them, many combinations may be made, which resemble silver entirely, and are more economical. The metal made according to the above proportions contains from 20 to 25 per cent. of silver, and corresponds inversely to the second standard alloy of silver, which is composed of 20 per cent. of alloy, with 80 per cent. of pure silver.

The metals employed should be of the best quality. The impure nickel is dissolved in muriatic, nitric, or diluted sulphuric acid. Chlorine is passed through the solution, and then the iron of the impure nickel is precipitated by ebullition with carbonate of lime.

The nickel is afterwards precipitated by carbonate of soda, dissolved again in hydrochloric acid, and the solution is diluted with a great quantity of water. After saturation by chlorine, an excess of carbonate of baryta is added to the solution, which is then allowed to rest. The nickel is afterwards precipitated in the metallic state by a galvanic current, or in the state of oxide, which is reduced in the ordinary way.

It is advantageous to melt the copper and the granulated nickel first, then to introduce the silver. A flux is employed, which is composed of borax and charcoal-dust. The ingots are rendered malleable by annealing them slowly and for a long time in charcoal-dust.

The employment of nickel on a large scale for white alloys dates back only a few years; at present it is an essential base of the compounds which are to be silvered.

The alloys known under the name of *maillechort,** sometimes, and wrongly, *melchior,* are made in France in the following proportions:—

Maillechort, first quality:—

Copper	8
Nickel	4
Zinc	3
	15

Second quality:—

Copper	8
Nickel	3
Zinc	3.5
	14.5

* Maillechort, German silver, argentan, and packfong are so much alike, that they may be considered as synonyms.—*Trans.*

Third quality:—

Copper	8
Nickel	4
Zinc	4
	16

More complex maillechorts have been made, but are not in great use, such as:—

Copper	55
Zinc	17
Nickel	23
Iron	3
Tin	2
	100

These proportions were those of the first composition of maillechort, patented more than thirty years ago.

We find in the trade several kinds of maillechort, more or less employed, under the name of:—

Paris Maillechort.

Copper	65
Nickel	16.8
Zinc	13
Iron	3.4
	98.2

German Maillechort.

Copper	50
Nickel	18.7
Zinc	31.3
	100.0

Chinese Maillechort.

Copper	50
Nickel	25
Zinc	25
	100

Maillechort for Spoons and Forks.

Copper	50
Nickel	20
Zinc	30
	100

Maillechort for Rolling.

Copper	60
Nickel	20
Zinc	20
	100

This last alloy may be subdivided into qualities, by varying the proportions in the same manner as we have indicated for the three qualities of maillechort.

The following alloys also belong to the class of maillechorts, argentans, German silver, &c.; that is to say, contain nickel as one of the principal bases:—

Electrum.

Copper	8
Nickel	4
Zinc	3.5
	15.5

This combination, which is nothing else but a maillechort of the first quality, imitates burnished silver, and is not so easily tarnished.

Tutenag.

Copper	8
Nickel	3
Zinc	5.5
	16.5

It is a maillechort of an inferior quality, which corresponds to the ordinary quality of the packfong, formerly imported from China. This alloy is very hard, difficult to be laminated, cannot be drawn out into wires, and is good for casting only.

The founders whose specialty is the manufacture of the alloys of copper with nickel and zinc, whether for maillechorts, or for similar products under different names, concur in admitting that the best alloy for beauty, lustre, &c., is made in the following proportions:—

Copper	8
Nickel	6
Zinc	3.5
	17.5

It is also the most costly among similar alloys, on account of the large proportion of nickel.

Alfénide is another compound which may be classified among the maillechorts, but those of a lower standard. It is well adapted for electro-silver-plating spoons, forks, and other articles with a smooth surface; but it does not succeed so well for decorated pieces, because the deposit of silver—and this is true of all the sorts of maillechort and German silver, to a greater or less degree—does not resist the fire, the acids, or the air as well as upon brass. The composition of alfénide is generally:—

Copper	60
Zinc	30
Nickel	10
Iron	1
	101

Let us now mention the alloy of Mr. Toucas, which may be added to the preceding compounds, and is made of—

Copper	5
Nickel	4
Antimony	1
Tin	1
Lead	1
Zinc	1
Iron	1
	14

This alloy has the advantage of being complex, if it does not possess other qualities than similar compounds. According to the inventor, it has nearly the color of silver, may be worked like it, and laminated by the ordinary processes. It is resisting, malleable, susceptible of a fine polish, with the lustre of platinum, and may be silvered perfectly well. For objects which are to be spun, hammered, or chased, the above alloy is convenient; but for cast and adjusted pieces it is preferable to increase the proportion of zinc, in order to increase the fluidity of the metal. This compound is employed for ornaments, jewelry, harness, etc.

Besides the nickel, which is used to impart to the alloys for false silverware the required hardness, whiteness, sonorousness, &c., manufacturers employ the alloys of copper, zinc, tin, lead, and sometimes antimony, bismuth, and arsenic, for white compounds, which to a certain point possess the qualities of the preceding alloys.

We here give a few of these compounds:—

English Tutania (white metal).

Copper 7 } Zinc 3 }	Brass	12
	Tin	12
	Bismuth	12
	Antimony	12
		48

The bismuth and the antimony are added to the molten alloy of brass and tin. The proportion of the brittle metals may be varied until the alloy has acquired the desired hardness and color.

German Tutania (white metal).

Copper	0.4
Tin	3.2
Antimony	42.0
	45.6

Spanish Tutania (white metal).

Iron and steel scraps	24
Antimony	48
Nitre	9
	81

The iron and steel must be heated to whiteness, and the antimony and nitre gradually added. 60 grammes of this composition are combined with 480 grammes of tin (about 2 oz. to 1 pound), in order to finish the alloy. A small proportion of arsenic is said to produce a very fine white metal.

Engestrum Tutania.

Copper	4
Antimony	8
Bismuth	1
	13

This compound, added to 100 parts of tin, produces a white metal which is employed in England for the manufacture of certain table wares. The following metals are also used in the same country, under the name of *Queen's metal*, for the manufacture of teapots and other vases imitating silver:—

Tin	3 to 9
Antimony	1 " 1
Bismuth	1 " 1
Lead	1 " 1
	6 to 12 parts.

The proportion of tin alone varies.

Another:—

Copper	2
Tin	50
Antimony	4
Bismuth	0.5
	56.5

Or:—

Copper 7 } Zinc 3 } Brass	24	Antimony	. .	0.8
Antimony	96	Bismuth	. .	18
Tin .	30	Lead	. . .	32

In France similar compounds are known under the names of *Algiers metal, minofor*, and *metal argentin.* Their usual composition is:—

Algiers Metal.

I.		II.	
Tin . . .	90	Tin . . .	94.5
Antimony . .	10	Copper . . .	5
		Antimony . .	0.5
	100		100.0

The alloy I. is for the manufacture of spoons, forks, goblets, &c.; it has been, or is yet, employed for plates for engraving music. It is capable of acquiring a very handsome polish.

The alloy II. is more especially employed for small hand-bells.

Metal Argentin (silver-like metal).

Tin	85.5
Antimony	14.5
	100.0

This alloy, as the Algiers metal No. I., is employed for making forks and spoons.

The following metal is used for coffeepots, teapots, and all similar vases:—

Minofor.

Copper	3.25
Tin	67.50
Antimony	17
Zinc	8.95
	96.70

The various white alloys which we have just indicated may be classified among the name of *Britannia metals*, which, at the present time, are very much sought for on account of their fluidity and their facility of acquiring a brilliant polish. The consumption of Britannia metal is considerable in England for low-priced wares.

The composition of these alloys is exceedingly variable, and we shall confine ourselves to the indication of the principal combinations.

As a rule, the preparation of these compounds is based on the idea of rendering tin harder, tougher, more sonorous, and more easily polished.

Copper and antimony impart to it these qualities; but, and as regards antimony, its proportion must not be exaggerated. An excess of antimony will not only impair the malleability of the alloy, but may also be dangerous to the health, as antimony is considered a poisonous metal, which does not resist the action of the vegetable acids.

Britannia metal will furnish castings as fine and sharp as those made with the most fluid alloys of tin and lead, copper and zinc, &c. It acquires a finer polish than the alloys of tin and lead, whereas the latter is too soft to bear the action of emery and other polishing materials.

All these advantages cause Britannia metal to rank among the most useful alloys.*

The most simple formula of *Britannia metal* is—

Tin	9
Antimony	1
	10

which is equally suitable for casting and rolling.

For similar alloys copper and zinc are employed in the following proportions:—

* For all the alloys of tin and copper, where tin largely predominates, it is better to have prepared, in advance, an alloy of tin and copper, rich in copper, which is called a *temper*, and is added to the definitive alloy in the proportion desired. By doing so, the alloy is more homogeneous, and there is less waste by oxidation, as the point of fusion is not very high.—*Trans.*

Tin	85 to 90
Antimony	5 " 10
Zinc	0.5 " 2
Copper	1 " 3

Bismuth is added to other alloys, and an alloy has been made of—

Tin	85
Antimony	5
Bismuth	5
Zinc	1.5
Copper	3 5
	100.0

Plate pewter belongs to the Britannia alloys, and is, as its name indicates, especially intended for rolling. Its composition is—

Tin	90
Antimony	7
Bismuth	2
Copper	2
	101

Certain kinds of Britannia contain neither zinc nor bismuth. Such is the *Ashberry metal*, made of—

Tin	78 to 82
Antimony	16 " 20
Copper	2 " 3

When we adopt the alloy made of the five metals tin, antimony, bismuth, zinc, and copper, we may employ the following proportions:—

1 part of brass (copper and zinc) made in advance,
1 " tin,
1 " bismuth,
1 " antimony,

which are melted together, and then remelted. During this last operation, from 15 to 20 per cent. of tin is added, according to the judgment of the manufacturer.

A more complex alloy, called *English metal*, is formed of—

	Tin	88
	Pure copper	2
Copper 75, Zinc 25	Brass	2
	Nickel	2
	Bismuth	1
	Antimony	8
	Tungsten	2

Mr. Karmarsch, who has thoroughly studied the properties of the Britannia alloys, says that the specific gravity of the alloys is 7.339 for laminated sheets and 7.361 for castings. He explains this anomaly by the fact that the molecules, under the action of the rollers, have a tendency to become separated, their softness and malleability not being great enough to allow of a regular and uniform compression. This is not an isolated fact. M. Le Brun has also found a lower specific gravity for certain alloys of copper and zinc, which had been laminated or hammered.

Certain Britannia alloys are very elastic, and well fitted for making wire. In this respect, they possess nearly the same amount of tenacity as pure tin.

Britannia metal is easily stamped and laminated, although it has a tendency to break under the rollers.

The casting is generally performed in metallic moulds of cast iron or brass. The different parts, for instance the feet and the handles of teapots, are soldered together with tin. The polishing is effected with fine sand and dry tripoli.

A great many articles of Britannia metal are, at the present time, silvered by the galvanic process, the same as other objects of German silver, Chinese packfong, or maillechort, which are so well manufactured in England, France, and Germany, that it is difficult to distinguish them from pure silver.

In some cases the Britannia metal is covered, by galvanism, with a deposit of tombac.

A small addition of a solution of gold to the bath of

copper and zinc, imparts to the deposit the color of similor.

The Britannia alloys and the analogous compounds which require bismuth or antimony, and nickel occasionally, ought to be classified among the common white metals, rather than among the metals of a certain value. But as these alloys are employed for articles of luxury, where they are made into artistical patterns, we have thought it better to separate them from the more common white compounds made only with tin, lead, or zinc, and to give them a place in this chapter.

For the same reason we shall mention a few more alloys, of which platinum is a component part, and which properly belong to those trades where the finish imparted to the work corresponds with the value of the metals employed.

Mock Gold, or False Gold.

Copper	16
Platinum	7
Zinc	1
	24

Ductile Alloy of Gold with Platinum.

Pure gold	30
Platinum	2
	32

The platinum is to be added only when the gold is in perfect fusion. The two combined metals give an alloy which is of a lighter color than pure gold, more fusible, and very ductile and elastic. These qualities may be found useful for certain works, especially for delicate springs, which cannot be made of steel.

The alloys of gold and platinum have been studied by an English savant, Mr. Prinsep, with a view of estimating the temperatures of blast-furnaces, and other apparatus where a powerful heat is employed. But these experiments have not given better results than that previously obtained with platinum alone.

Alloy for mirrors, ductile, notwithstanding its hardness, unalterable in the air, and receiving a brilliant polish:—

Platinum	60
Copper	40
	100

Metals for Cutlery.

Steel alloyed with $\frac{1}{500}$ of platinum or silver, which is harder and more malleable than steel alone.

Also steel with rhodium, &c. &c.

X.

WHITE ALLOYS.

We include in this category all the alloys which are not used in the manufacture of what may be called articles of luxury, and which have not been mentioned in the preceding chapter.

These alloys, of which we shall indicate the combinations most employed in the arts, are very important, as will be seen.

The alloys of zinc, tin, and lead, which have already been studied in the second part of this book, may, in certain proportions, furnish white metals which, if they do not present all the qualities, possess at least some of the characteristics, of the alloys called tutania, queen's metal, German silver, minofor, Britannia metal, &c.

The ternary alloys of zinc, tin, and lead are more economical than the former combinations, do not tarnish more, are as easily polished, and may be laminated. The best proportions are within these limits:—

Tin	16	16
Zinc	4	3
Lead	4	3

It is proper to melt the zinc at the lowest temperature possible, to add tin, and then lead. The whole is carefully stirred, and the bath is covered with borax and charcoal-dust, or rosin, in order to prevent oxidation. The proportion of zinc is increased, if toughness and hardness are desired; more tin increases the malleability, the whiteness, and the polish; but the proportion of lead should not be much greater than those indicated above.

To these metals we sometimes add copper, antimony, or bismuth, in order to obtain the following compounds:—*

English Alloys for Casts from Engravings, Stereotypes, &c.

No. 1. Common quality.

Tin	3.36
Lead	0.48
Copper	0.18
Zinc	0.60

No. 2. Ordinary quality.

Tin	100
Antimony	17

This quality belongs to the series of the alloys for type-founders, the same as the following ones, which have already been indicated, or have nearly the same composition:—

Lead	9
Antimony	2
Bismuth	1

* The white metals, which are not classified here, will be found elsewhere. The alloys which form this chapter are those which we have not been able to classify under the various titles we have hitherto adopted.

Lead	10
Antimony	2

Lead	8
Antimony	2
Tin	1

No. 3. Superior quality.

Tin	5.76
Antimony	0.48
Copper	0.12

The copper must be melted first, and the other metals are added in the following order: tin and antimony.

Pewter is generally composed of—

Tin	80
Lead	20
	100

but gives its name also to the above alloy No. 2 (tin 100, antimony 17), and is then a pewter of first quality. According to Mr. Mackenzie, these proportions form the best combination of lead and antimony, as regards hardness, resistance, and whiteness.

The pewters are employed in England for the same uses as the French alloys, whose composition varies between—

Tin	82	92
Lead	18	8
	100	100

for common pots and plates.

Better articles, under the name of *Algiers metal*, are made of—

Tin	75	90
Antimony	25	10
	100	100

An *alloy improper for domestic uses* has been made of—

Tin	10
Steel filings	2
Metallic arsenic	1.5
Arsenious acid	2.5
	16.0

This alloy gives a white metal, ductile, malleable, and very easily cast. But its poisonous nature prevents it from becoming extensively used, except in some particular cases.

Alloy for Seats of Stopcocks.

Tin	86
Antimony	14
	100

This alloy retains its polish quite well, even in a damp atmosphere. According to Thénard, it presents the remarkable property that when it is dissolved in diluted muriatic acid, the two metals become precipitated.

Alloy for Plugs of Stopcocks.

Tin	80
Antimony	20
	100

This is harder and resists friction better than the preceding.

Alloy for Keys of Flutes, Clarionets, &c.

Lead	20
Antimony	40
	60

This alloy is hard, and its polish is not easily tarnished.

Hard Tin.

Tin	1
Antimony	0.5

This alloy appears to be on the extreme limit of the alloys of tin and antimony which may be used.

Kustitien Metal for Tinning.

Tin	11.52
Iron	0.48
Antimony	0.15
	12.15

This alloy has a blue tint when polished. It is very good for tinning the insides of kitchen utensils made of wrought iron.

English Hard White Metal (common).

Copper 3, Zinc 1	Brass	480
	Zinc	45
	Tin	15
		540

Mock Platinum, or *False Platinum.*

Copper 3, Zinc 1	Brass	240
	Zinc	150
		390

Imitation of silver, especially as to its sonorousness:

Copper	448
Zinc	22
	470

White Metal, called *Prince's Metal.*

Copper, Bismuth } Variable proportions.

All these alloys are brittle. They present no other interest except their white color and their fine polish.

White Copper, or *White Tombac.*

Copper	75
Tin	25
	100

This metal is employed in England for the manufacture of buttons and small articles of hardware. Being sonorous, it may be used for hand-bells, &c.

Various alloys for buttons employed in England:—

No. 1. Superior quality.

Copper 3, Zinc 1	Brass	373
	Zinc	62
	Tin	31
		466

No. 2. Ordinary quality.*

Copper 3, Zinc 1	Brass	373
	Zinc	47
	Tin	47
		467

No. 3. Common quality.

Copper 3, Zinc 1	Brass	373
	Zinc	140
		513

Vogel's alloy for polishing steel is employed in the shape of thin blades or files for applying rouge to the small pieces of steel of the watchmakers, and is composed of—

Copper	8
Tin	2
Zinc	1
Lead	1

This alloy, which we have studied in the quaternary combinations of copper, tin, zinc, and lead, is very hard, resists the tools, and must be ground upon a stone.

* From its composition, there being more tin and less zinc, No. 2 appears to be the superior quality, and No. 1 the ordinary quality. —*Trans.*

XI.

FUSIBLE ALLOYS.

This name is applied to those alloys which are combined in such a manner that they will melt at a given temperature.

Although it is difficult to determine with perfect exactness their points of fusion, these fusible alloys may be useful in the arts and in manufactures for ascertaining a given temperature; for obtaining plastic metals easily melted, in order to obtain casts of delicate objects which may be damaged by too high a temperature; for making very fusible soft solders; and lastly, as a matter of precaution for such apparatus as is liable to be instantaneously destroyed by a sudden and excessive increase of temperature. In this latter connection may be named the fusible safety plates or plugs of boilers.

These safety plates were at the beginning very extensively used; but at the present day they are rarely to be met with, and are no longer required by the rules which regulate boilers and steam-engines. However, it may be found useful to know the composition of these alloys.

The fusible alloys are based on the property of certain metals to become more fusible when combined, than they were when taken singly. Bismuth, tin, and lead, especially, follow this rule.

It is difficult to obtain these alloys in a perfectly homogeneous state. They have a tendency to become decomposed while yet in a state of fusion, the lead going to the bottom of the fused mass.

The *alloy of Darcet* or *of Rose* is made of—

Bismuth	50
Tin	30
Lead	20
	100

and is fusible at 100° C. (boiling water). A peculiarity of this alloy is, that it will become hot again, and enough to burn the fingers, after it has been cooled in cold water. The cause of this phenomenon is, that during the solidification and crystallization of the inside portions of the alloy, the latent heat of these parts is immediately transmitted to the cooled surface.

Mr. Darcet indicates the following alloys, which result from his own experiments, and the proportions of which are:—

No. 1. Bismuth 70, lead 20, tin 40.—Softens at 100° C., without melting, and may be kneaded in the fingers.

No. 2. Bismuth 80, lead 20, tin 60.—Softens at 100° C., and is easily oxidized. There is, however, too much tin.

No. 3. Bismuth 80, lead 20, tin 40.

No. 4. Bismuth 160, lead 40, tin 70.

No. 5. Bismuth 90, lead 20, tin 40.

These three alloys become more or less soft at 100°. No. 4 becomes softer than either No. 3 or No. 5.

No. 6. Bismuth 160, lead 50, tin 70.—Becomes nearly fluid at 100°.

No. 7. Bismuth 80, lead 30, tin 40.—Becomes liquid at 100°; but not very fluid.

No. 8. Bismuth 80, lead 40, tin 40.—Very liquid at 100°.

No. 9. Bismuth 80, lead 70, tin 10.—Becomes soft at 100°, but does not melt.

No. 10. Bismuth 160, lead 150, tin 10.—Neither liquid nor soft at 100°.

These alloys are generally harsh; nevertheless, they may be cut. Their fracture is a dead blackish-gray. They are rapidly tarnished in the air, and more so in boiling water, in which they become covered with a wrinkled pellicle, which falls as a black powder.

A few savans have studied with great persistency the *fusible combinations of bismuth, lead, and tin.* The following table, made by MM. S. Parker and Martin, indicates the various points of fusion of these alloys:—

Metals of the Alloys.			Temperatures of fusion.	Metals of the Alloys.			Temperatures of fusion.
Bismuth.	Lead.	Tin.		Bismuth.	Lead.	Tin.	
Parts.	Parts.	Parts.	Degrees centigrade.	Parts.	Parts.	Parts.	Degrees centigrade.
8	5	3	202	8	16	24	316
8	6	3	208	8	18	24	312
8	8	3	226	8	20	24	310
8	8	4	236	8	22	24	308
8	8	6	243	8	24	24	310
8	8	8	254	8	26	24	320
8	10	8	266	8	28	24	330
8	12	8	270	8	30	24	342
8	16	8	300	8	32	24	352
8	16	10	304	8	32	28	332
8	16	12	290	8	32	30	328
8	16	14	390	8	32	32	320
8	16	16	292	8	32	34	318
8	16	18	298	8	32	36	320
8	16	20	304	8	32	38	322
8	16	22	312	8	32	40	324

MM. Parker and Martin have employed these alloys as metallic baths for tempering tools. It is possible in this manner to determine exactly the temperature best adapted for various cutting instruments.

The *alloys of lead and bismuth* have also been tried. They are too easily oxidized, and are difficult to make, on account of the separation of the lead. Bismuth increases the tenacity of lead. An alloy of equal parts of bismuth and lead possesses a tenacity from fifteen to twenty times that of pure lead.

The *alloys of bismuth and tin* succeed better. Those which are best known are—

Bismuth 50	Tin 50	Melting at about 160° C.
" 33	" 67	" " 166
" 10	" 80	" " 200

The *alloys of bismuth, lead, and zinc* have been but little studied. An alloy of equal parts of these three metals is fusible at about 100° C.

An *amalgam of lead, bismuth, and mercury*—

Lead	20
Bismuth	20
Mercury	60
	100

is very fluid at the ordinary temperature, and may be squeezed through chamois leather the same as pure mercury. This combination is sometimes employed for falsifying mercury; but, notwithstanding its fluidity, the drops, when made to run, have an elongated form.

Mr. Mackenzie indicates an *alloy fusible by friction*, which is a combination of 2 parts of bismuth melted with 4 parts of lead, and then thrown into a crucible containing mercury. This amalgam becomes solid by cooling, but if we break it, and rub the two portions against each other, they soon melt.

In general, the fusible compounds of bismuth, tin, and lead have their fusibility increased by the addition of mercury.

A *very fusible alloy for casts* is made by adding in weight a sixteenth of mercury to the already mentioned alloy, fusible at 100° C., and known as the Darcet or Rose alloy. The new compound is fusible at the temperature of the human body.

This quaternary alloy may be employed for obtaining casts of certain portions of the human body after death; the ear, for instance. The animal substances are destroyed by a concentrated solution of caustic potassa, and the metal remains.

An *alloy for silvering glass globes*, by means of a small pellicle deposited on the inside surface, is made of—

Bismuth	2
Tin	1
Lead	1
Mercury	10

An *alloy for fusible teaspoons, &c.*, is composed of—

Bismuth	8
Tin	3
Lead	5
Mercury	1 or 2

and is employed by amateurs in making amusing experiments with tea or coffee spoons, which immediately melt when plunged into a hot liquid.

Leaving aside bismuth, the arts employ *other fusible alloys*, among which we may notice the following ones:—

Tin 3 parts, lead 2 parts. Fusible at 167° C.

Lead 4 parts, antimony 1 part. Fusible at a red heat, or about 500° C.

Lead 1 part, zinc 1 part. A very tenacious compound, resisting friction well, has a brilliant lustre, is hard, somewhat ductile, and melts at a temperature varying from 460° to 500° C.

Tin 2 parts, zinc 4 parts. Melts between 300° and 350° C.

Tin 3 parts, zinc 4 parts. Melts between 320° and 360° C.

Tin 1 part, zinc 3 parts. Melts between 280° and 300° C.

We now pass to the *Appold alloys*, useful for ascertaining certain given temperatures. The principal of these alloys which were composed by MM. Appold Brothers, in order to determine the temperature of their apparatus for making coke, are:—

Copper 4	Tin 1	Melting at about 1050° C.
" 5	" 1	" " 1100
" 6	" 1	" " 1130
" 8	" 1	" " 1160
" 12	" 1	" " 1230
" 20	" 1	" " 1300

In this connection we may state that the majority of alloys may be employed, in certain cases, as fusible alloys. It is sufficient to carefully determine the point of fusion of the alloys with proper instruments, and then to construct methodical tables in which are recorded the variations of temperature corresponding to the nature of the alloys employed, and the proportions of the component metals.

XII.

ALLOYS FOR MACHINERY, ANTI-FRICTION METALS, &c.

We classify these alloys in three distinct categories:—

Bronze alloys.

Brass alloys.

White alloys.

Bronze alloys are employed by the constructors of machinery wherever certain conditions of tenacity, wear, hardness, and resistance to friction are required. The following are extensively used:—

Bronze for pumps, pillow blocks, nuts &c.:—

Copper	88
Tin	12
	100

The same, but harder:—*

Copper	90
Tin	10
	100

These bronzes are employed in the government shops and other large works. An addition of from 1 to 4 parts of zinc is allowed in certain cases.

Alloys for blocks of connecting rods and collars for eccentrics:—

* We should suppose that the proportion of tin being smaller, this alloy would be softer than the preceding.—*Trans.*

Copper	83	Copper	83
Tin	15	Tin	15
Zinc	2	Zinc	1.5
		Lead	0.5
	100		100.0

Or—

Copper	84	Copper	84
Tin	14	Tin	14
Zinc	1.5	Zinc	2
Lead	0.5		
	100.0		100

if the alloy is desired slightly softer and more malleable.

The following *alloys for journals of locomotive driving axles* are employed by English makers:—

Copper	74
Tin	9.5
Zinc	9.5
Lead	7
	100.0

Others are satisfied with—

Copper	80	85.25
Tin	18	12.75
Zinc	2	2
	100	100.00

Alloys for blocks with collars of connecting rods, which require a milder and more malleable metal:—

Copper	82
Tin	16
Zinc	2
	100

Bronze for pistons:—

Copper	89.75
Tin	2.25
Zinc	8
	100.00

Alloy for locomotive axle-journals:—

Copper	80
Tin	18
Zinc	2
	100

Or—

Copper	79
Tin	18
Zinc	2.5
Lead	0.5
	100.0

Alloy for journals of cranes, winches, &c., as required by the Northern Railway of France for the apparatus of its fixed stock:—

Copper	82
Tin	18
	100

Alloy for journals of wagons employed by the same company :—

Copper	86
Tin	14
	100

We see that all these bronzes have very much the same composition. The proportion of copper is rarely below 80 per cent., and that of zinc ranges between 2 and 3 per cent.

A slight variation in the proportions of the alloy may be noticed in practice. This explains why we have indicated the principal combinations in daily use, although several of them differ very little from each other. For the same reason we shall notice the following alloys :—

Alloy for locomotive whistles:—

I. A clear sound, for passenger engines—

Copper	80
Tin	18
Antimony	2
	100

II. A deeper pitch, for merchandise machines—

Copper	81
Tin	17
Antimony	2
	100

Mild alloy for pumps, clappers or valves, and stop-cocks:—

Copper	88
Tin	10
Zinc	1.75
Lead	0.25
	100.00

Or—

Copper	88
Tin	10
Zinc	2
	100

Bronze for ball valves and pieces to be brazed:—

Copper	87
Tin	12
Antimony	1
	100

Alloy for cleaning plugs:—

Copper	98
Tin	2
	100

This composition may be forged like pure copper, for which it is a substitute. The addition of tin renders the casting more easy and sound.

Hard alloy for bearings of merchandise and ballast wagons:—

Copper	78
Tin	20
Zinc	2
	100

The next composition has been tried for the same purpose, but without advantage:—

Cast iron	70
Copper	25
Zinc	5
	100

The following alloys are employed at the important works of Seraing for Belgian locomotives. Their composition is very nearly that of the corresponding alloys which we have already mentioned.

Bronze for journals of locomotive driving axles:—

Copper	86
Tin	14
	100

Copper	89
Tin	8
Zinc	3
	100

Bronze for blocks of side valve connecting rods:—

Copper	85.25
Tin	12.75
Zinc	2.00
	100.00

Bronze for regulators:—

Copper	86.82
Tin	12.38
Zinc	0.80
	100.00

Bronze for stuffing boxes:—

Copper	90.25
Tin	3.50
Zinc	6.25
	100.00

Bronze for pistons:—

Copper	89
Tin	2.5
Zinc	8.5
	100.0

The alloys of brass are employed in mechanical constructions when the resistance of the metal is not exposed to very great strains, and for economical or ornamental purposes.

The brasses for machinery generally have a composition ranging from 20 to 35 per cent. of zinc, and from 80 to 65 per cent. of copper. With less than 20 parts of zinc, the alloy becomes red, and may be applied to some particular purposes; but it is no longer to be considered as brass. With more than 35 parts of zinc, the alloy is harsh, brittle, and whitish; and, although it may be employed for certain common uses, it is no longer a brass for mechanical purposes.

The brass compounds most generally employed in the arts are:—

Brass for turners:—

Copper	61.6
Zinc	35.3
Tin	0.5
Lead	2.5
	99.9

Or the three following compositions, presenting various shades:—

No. 1.—	Copper	79.5
	Zinc	20
	Lead	0.5
		100.0

No. 2.—Copper	74.5
Zinc	25
Lead	0.5
	100.0

No. 3.—Copper	66.5
Zinc	33
Lead	0.5
	100.0

The *brass employed in the French navy*, and in the *Ecoles des Arts et Métiers*, is generally made as follows:—

Copper	65.80
Zinc	31.80
Tin	0.25
Lead	2.80
	100.65

This alloy, when polished, has a pleasing greenish-yellow color, and is quite malleable. It is especially employed for large pieces of machinery.

The *brass for small pieces of machinery* is of another composition, as follows:—

Copper	76
Zinc	24
Lead	0.5
	100.5

Brass for thin pieces, hinges, &c.:—

Copper	85
Zinc	15
Lead	1
	101

Several English railways have employed for the journal boxes of locomotive and wagon axles the *Fenton alloys*, which are intermediate between the bronzes and the brasses. Alloy No. 1 has given quite good results.

No. 1.—Copper	56
Zinc	28
Tin	16
	100

This compound appears to resist friction well without much heating, and its specific gravity is below that of the ordinary bronzes. It corresponds to the combination made by Margraff in his experiments on the alloys of copper, tin, and zinc, and which was made of copper 100, tin 50, and zinc 25 parts. The metal obtained by this chemist was of a yellowish-white color, with an irregular grain, very hard, although quite easily filed, but without any malleability.

No. 2.—Copper	5.5
Zinc	80.0
Tin	14.5
	100.0

This alloy is more advantageous than the preceding as regards economy and lightness. It has been employed not only for journals which, it has been said, required but little oiling, but also for many kinds of pieces submitted to friction, stuffing-boxes, valves, slide bars, &c.

These alloys, notwithstanding their qualities, which appear to have been exaggerated, are difficult to make. They are not directly made in one operation, but as follows: The pure copper is melted in a crucible, to which is added a brass composed of copper 70, and zinc 30, and then the tin. When all is melted and well stirred, it is cast into ingots, which constitute *hard metal.*

For producing the definitive alloy the zinc is melted in a crucible, and the hard metal, previously melted in another crucible, is poured into it. It is thoroughly mixed, and a new proportion of tin may be added, according to the degree of hardness or softness required.

Before casting, the metal is again stirred. The alloy, especially during the melting of the zinc, ought to be covered with a thick layer of charcoal dust, in order to avoid the loss by volatilization or oxidation.

The Fenton alloys, and all similar compounds, appear to be, as antifriction metals, intermediate between the bronzes and the white metals. The latter have for a certain length of time been much employed by constructors who regarded them as very economical in first cost and in lubricating principles.

The white alloys have been experimented upon especially for lining the journal boxes of locomotive and wagon axles; but we believe that everywhere, after having tried the bronzes and the white alloys in comparison, the former have been found more advantageous, as they last longer and are not so easily scratched by the dust as the white alloys.

Mr. Nozo, the skilful engineer of the repair shops of the Northern Railroad, has published in the *Bulletins de la Société des Ingénieurs Civils* the results of his experiments on antifriction metals, and has condemned the white metals, even those which had been the most extolled, such as *Grafton's antifriction metal, Vaucher's metal, Detourbet's metal, &c.* The conclusions of Mr. Nozo are:—

That the white metals, whether for whole journals or their linings, may be advantageously employed in machinery revolving with a small velocity, or with an average velocity and small strain; but that they are not suited to the rolling stock of railroads in which the strains and the velocity are such as to rapidly wear all the metals which are not hard enough to resist an energetic friction.

We will now mention a few of the white alloys at present in use:—

No. 1. *White alloy for lining journal boxes, collars, pillow blocks, &c.:*—

Copper	4
Tin	96
Antimony	8
	108

12 parts of copper are melted, to which are added 36 parts of tin, then 24 parts of antimony, and lastly 36 parts of tin. As soon as the copper is melted the temperature is lowered in order to prevent the oxidization of the tin and antimony, and the surface of the bath is protected from the contact of the air. The first composition, made as aforesaid, is employed for the definitive alloy, which is made of 50 parts of the first alloy and 100 parts of tin.

The pieces of machinery which require only a lining are luted with clay, and the melted alloy is poured into its proper place, with enough metal to compensate for the shrinkage.

No. 2. *White alloy for small journals*, and when the friction is not very great:—

Copper	9
Tin	73
Antimony	18
	100

This alloy may be polished with dry materials, and wears well. It would be more economical if a small proportion of lead were added, but its resistance and durability would be impaired.

No. 3. *White alloy for bearings*, made on the same principles as the preceding ones:—

Copper	1
Tin	50
Antimony	5
	56

This alloy is more economical and has a more greasy

touch than compositions No. 1 and No. 2. It is very good for machines which are not overworked.

No. 4. *White alloy to be cast directly in journal boxes:—*

Lead	32
Zinc	18
Antimony	50
	100

No. 5. *Soft alloy for pillow blocks:—*

Lead	85
Antimony	15
	100

This alloy, which may also be cast directly in its place, becomes heated with difficulty, and is said to resist well a rapid friction.

A similar but more complete alloy is *Vaucher's alloy.* It has been extensively employed for lining the journal boxes of carriage and wagon axles, but is now nearly forgotten. Its composition is:—

Zinc	75
Tin	18
Lead	4.5
Antimony	2.5
	100.0

The zinc is melted first, then the tin and the lead are added. The antimony, which requires a greater heat, is melted separately and poured the last into the bath of zinc, tin, and lead.

This melted alloy is run through small vents or apertures, left at the upper part of the axle boxes; and small discs of sheet-iron at both ends of these boxes prevent the metal from escaping. In order to leave room for the lubricating material two or three turns of a thick ribbon are wound around the middle of the axle journal, and therefore the alloy does not reach these parts.

Vaucher's metal, which does not seem to us to possess any special qualities beyond the majority of the antifriction white metals, has been more or less imitated. Is it possible to admit patent rights on alloys? Among the imitations we have already cited are Détourbet and Grafton's metals, and we may add the *alloys of Goldsmith* and of *Dewrance,* the latter being composed of 4 parts of copper, 8 of antimony, and 6 of tin. All these alloys are neither worse nor better.

A few years ago the *antifriction metals of Morries-Stirling* and *of Muntz* were extensively employed in England, and had in their composition a certain proportion of wrought or cast-iron, besides copper, tin, and zinc. These alloys were very irregular in their composition, and we do not believe that they have been employed in the French foundries, except in an experimental way.

The alloys prepared by Mr. Stirling, and tried in the arsenals of Woolwich, Portsmouth, and Chatham, had a resistance to flexion much greater than that of ordinary bronzes. Thus, the *bronze made at Woolwich,* in the following proportions, corresponding to various uses:—

Copper	20
Tin	2
Zinc	1
	23 parts

Copper	6	7	8	10
Tin	1	1	1	1

have shown an average resistance of 11.66 tons per square inch, while the resistance of the corresponding Stirling alloys was 16.42 tons on an average.

Again, bars one inch square and three feet long were placed upon supports 2 feet 3 inches apart. A load placed in their middle produced a deflection of 73.44 with the bronze of Portsmouth (copper 10, tin 1); while with the Stirling metal the deflection was only 16.79.

But notwithstanding these results the Stirling metal, which is difficult to obtain in a sound and homogeneous state, did not succeed.

Before Mr. Stirling's patent another metal, known as *Fazie metal*, from the name of its inventor, was patented in England, and composed of wrought-iron, cast-iron, and brass. These alloys were claimed to be more tenacious and to wear better than either of the component metals taken singly. The bronze or brass and the iron and cast-iron were melted separately, then mixed, and the stirring continued all the time, even when being poured out.

Karsten repeated these experiments by mixing with cast-iron a small proportion of copper, which had the effect of rendering the mixture less easily oxidized, but nothing has been gained from these experiments for ordinary practice in foundries.

XIII.

SOLDERS.

We shall mention two kinds of solders:—

1. The solders made by the fusion of the metal itself, without any other metals. These solders are possible with the majority of metals, even the refractory ones, cast-iron, for instance. We have spoken in one of our works of the processes of the *Autogenous solders*, but which do not find their place here, the subject being alloys.

2. The solders made upon a metal with another metal, or by an alloy applied to the surfaces which are to be united.

In the latter case the metal or the alloy must be more fusible than the metal to be soldered, and have for it a powerful chemical affinity.

In general the soldering is the more perfect as the point of fusion of the metal to be soldered and that of the soldering metal or alloy approach each other.

When the parts to be soldered and the solder may be brought to an incipient, or even a complete fusion, the maximum of resistance will be obtained, the solder having formed a true alloy with the soldered metal.

A *strong or hard solder* is employed for metals difficult to melt, and which, being soldered, have to resist the action of the heat. The *soft solders*, with a base of lead and tin, are much more fusible than the metals to be united, and are employed when great solidity is not required, and when they are not subjected to the action of heat.

For making copper solders the copper is melted in a crucible, and then the zinc, previously melted in another crucible, is added. The whole is thoroughly stirred, and when the alloy is at the proper temperature, it is poured from a certain height upon a bundle of birch twigs kept wet and agitated at the surface of a tub of water. The solder is thus obtained in the shape of fine grains having an irregular crystallization.

When this solder is not sufficiently fine or regular it is broken in a cast-iron mortar, and passed through a sieve.

The manufacturers of solder generally prefer to cast the hard solder into ingot moulds instead of using the above process, which is good enough for shops. The cooling is prevented as much as possible in order to develop the crystallization, which helps the subsequent operations of crushing and sifting.

The solders most generally employed in the arts are:—

Solders for Iron.

Pure granulated copper, or—

Copper	67
Zinc	33
	100

Or:—

Copper	60
Zinc	40
	100

The last two alloys, which may be replaced by a powdered brass holding from 33 to 40 per cent. of zinc, are also employed for small pieces of iron and copper.

Solders for pure copper and brass.

Hard Solder for Tubes of Pure Copper.

Copper	3
Zinc	1
	4

Or:—

Copper	7
Zinc	3
Tin	2
	12

Or, a brass containing 70 parts of copper to 30 of zinc; or 75 of copper to 25 of zinc.

Middling hard solder, more fusible than ordinary brass:—

Scraps from the metal to be soldered	4
Zinc	1
	5

The proportions generally admitted in the French navy yards are:—

Hard Solder for Small and Thin Pieces.

Pure copper	86.5
Zinc	9.5
Tin	4.
	100.0

This solder is a light yellow, with fine and quite regular grains similar to filings. It will become oxidized without melting, unless it is kept in the middle of the fire and thus melted rapidly.

The same alloy, but coarser, may be employed for soldering large pieces.

Middling Hard Solder for Small Pieces of Brass.

Copper 70, *Tin 30	Brass	69.5
	Zinc	18.5
	Tin	12.
		100.0

Middling Hard Solder for Tubes of Brass, or of Thin Copper.

Copper 70, *Tin 30	Brass	77.5
	Zinc	17.5
	Tin	5.
		100.0

Middling Hard Solder for Soldering the ends of Brass Tubes together, or to Flanges.

Copper 70, *Tin 30	Brass	77.5
	Zinc	20.5
	Tin	2.
		100.0

* The name of *cuivre jaune* or *laiton* (literally *yellow copper*, or *brass*), given by the author, implies the presence of zinc, instead of tin, in its composition. Although we retain the word tin in the foregoing and following alloys marked with the asterisk, we strongly incline to believe that it should be zinc.—TRANS.

Middling Hard Solder for uniting Brass Tubes along their lengths, and is to be preferred to the former compounds when the soldered portions are to be hammered afterwards:—

Copper 70	} Brass	. . . .	77.5
*Tin 30			
	Zinc	. . . .	22.5
			100.0

Other kinds of *solders for pure copper* are sometimes employed. They are alloys of copper and lead in various proportions, as for instance:—

Copper . . .	100	Lead . . .	25
" . . .	100	" . . .	20
" . . .	100	" . . .	18
" . . .	100	" . . .	16

These alloys are sufficiently fusible, have the color of copper, and may be used for brazing it, without borax. They are malleable, clog the file, and are quite serviceable as a solder. To prepare them the copper is melted first, then the molten lead is added to it, just before pouring out. These solders are granulated by the ordinary processes.

SOFT SOLDERS.

Among the soft solders to be employed with metals melting at a low temperature, we may notice the following ones:—

Solder for Plumbers.

Lead	1 or	2
Tin	1	1
	2	3

Soft Solder.

Lead	1
Tin	2
	3

Solder for Tinned Iron.

Lead	7
Tin	1
	8

Solder for Pewter.

Lead	1
Tin	2
	3

This solder, which is employed in England by the manufacturers of pewter wares, is the same as that known in France under the name of soft solder.

Alloy for Sealing up Iron in Stone.

Lead	2
Zinc	1
	3

This alloy is more resisting, and adheres better than pure lead.

It has been tried, in certain cases, to substitute the *zinc solders*, or amalgams of zinc, for the ordinary soft solders. When soldering with zinc, this metal is cut into thin strips and put with a flux between the edges of the metal to be soldered; or a granular amalgam of zinc is employed with an appropriate flux. The surfaces to be united are heated up until the zinc melts, and sometimes to redness, according to the metals employed. The fluxes are generally borax or sal ammoniac.

Soft solders of bismuth, tin, and lead are sometimes used, and their compositions will be found in the chapter on fusible alloys.

Solders for jewelry, silver or gold wares, ornaments, &c.

We employ the following solders for jewelry and the precious metals :—

Hard Solder for Gold.

Gold (18 carats or $\frac{750}{1000}$)	18
Silver	10
Pure copper	10
	38

This solder and the following ones are made with fine filings of the metals, which are melted together:—

Gold solder called one-fourth .	gold 3 alloy .	1
" " " *one-third* .	" 2 " .	1
" " " *one-half* .	" 1 " .	1

The alloy is made of 66 per cent. of pure silver, and 33 per cent. of copper, except for the solder "one-half," when the proportions are equal parts of silver and copper.

Hard Solder for Silver.

Silver	66
Copper	23
Zinc	10
	99

This solder is more fusible than the middling hard solders for copper, and is sometimes used for brazing brass:—

Silver solder called one-sixth silver . . 5 . .	brass . . .	1
" " *one-fourth* " . . 3 . .	" . . .	1
" " *one-third* " . . 2 . .	" . . .	1

In order to obtain a homogeneous product these solders ought to be melted several times. The metal is then laminated into thin bands, which are granulated into spangles, ready to be mixed with borax.

If a piece of silverware is to be soldered several times, it is proper to employ, at the beginning, the richer solders, which, being less fusible, will not be subject to displacement by the solders of lower standards, employed at the end of the operation.

Other silver-solders are employed, such as—

Silver	2
Bronze	1
	3 parts.

Silver	4	1
Bronze	3	1
Arsenic	0.25	1
	7.25 parts.	3 parts.

Silver	2
Dutch gold (brass)	1
Arsenic	0.5
	3.5 parts.

The arsenic is added to the bath after the fusion of the other metals. These various solders are drawn out under the hammer, or laminated and then cut into spangles.

Solder for Platinum.

Pure gold, or gold with ½ per cent. of an alloy of platinum and iridium.

Hard Solder for Aluminium Bronze.

Gold	88.88
Silver	4.68
Copper	6.44
	100.00

Middling Hard Solder for Aluminium Bronze.

Gold	54.4
Silver	27.
Copper	18.6
	100.0

Soft Solder for Aluminium Bronze.

Copper	70	Brass	14.3
Tin*	30		
		Gold	14.3
		Silver	57.1
		Copper	14.3
			100.0

Solder for German Silver.

Copper	8	German silver	5
Nickel	2		
Zinc	3.5		
		Zinc	4
			9

This alloy is cast into thin plates, which are cut and pulverized. Its texture has a dead lustre, and is slightly fibrous. It is the more ductile, as the proportion of zinc is smaller.

Silver solder for plated ware, employed in England:—

Pure silver	2
Bronze	1
	3

Amalgam of Copper.

Copper	30
Mercury	70
	100

XIV.

MISCELLANEOUS ALLOYS.

This last series comprises the alloys which we have not been able to classify in the preceding series. We here insert all such compounds that we have picked up from our own works, or from treatises on alloys.

A few of these compounds are really useful, while others will look very empirical. We give them as we

* See foot note page 262.—TRANS.

find them in the works of certain authors, who have tried or verified them no more than we have.

Alloys for small patterns in foundries:—

No. 1.—Tin	. . . .	7.5
Lead	. . . .	2.5
		10.0 parts.
No. 2.—Zinc	. . . .	75
Tin	. . . .	25
		100 parts.
No. 3.—Tin	. . . .	30
Lead	. . . .	70
		100 parts.

The last of these alloys is for patterns which will not be in frequent use, and which may be mended, bent, &c. The first gives harder and stiffer patterns; the second is harder than tin and more tenacious than zinc, at the same time that it preserves a certain ductility.

With from 15 to 20 per cent. of tin, the zinc becomes less brittle, and is better adapted to many useful purposes. With from 15 to 20 per cent. of tin, lead becomes harder and more resisting. Even from 2 to 5 per cent. of tin are sufficient to harden lead. On the other hand, a small proportion of lead renders tin more supple, easily worked, and not so subject to cracks.

An addition of bismuth to lead increases the hardness of the latter metal. The alloy which possesses the maximum of tenacity is about:

Lead	. . . .	60
Bismuth	. . . .	40
		100

PLASTIC ALLOYS.

The best alloys of lead, tin, and bismuth, for obtaining casts of medals, coins, &c., are comprised within the following proportions:—

No. 1.—*Krafft's alloy*:—

Bismuth	5
Lead	2
Tin	1
	8

This alloy is fusible at about 104° C.

No. 2.—*Homberg's alloy*:—

Bismuth	3
Lead	3
Tin	3
	9

This alloy is fusible at 122° C., has nearly the appearance of silver, and is quite hard. It is used in England for casts of medals.

No. 3.—*Alloy of Valentin Rose*:—

Bismuth	4 to 6
Lead	2 2
Tin	2 to 3
	8 to 11

This alloy melts between 100° and 130° C.

No. 4.—*Alloy of Rose* (the father):—

Bismuth	2
Lead	2
Tin	2
	6

which melts at 93° C.*

These alloys, of which the points of fusion may be quite accurately determined, have been tried for tempering cutting instruments.

The *martial regulus* is also employed for medals and objects in relief, and is composed of—

* It is curious to observe that the alloys Nos. 2 and 4, both made of equal parts of the same metals, melt at different temperatures. This probably depends on their homogeneousness.—TRANS.

Antimony	7
Iron	1
	8

According to certain authors, the casts are sharper than those of cast-iron.

The following metal, called *expansion metal*, possesses the property of expanding when cooling, and is therefore very useful for filling small defects in metallic pieces, and for sealing and obtaining certain casts:—

Lead	6
Antimony	2
Bismuth	1
	9

Various compounds.

An English author indicates the following *amalgam for varnishing plaster casts*:—

Tin	1
Bismuth	1
Mercury	1
	3

The mercury is added to the tin and bismuth already melted, and the whole is thoroughly stirred, in order to perfect the combination. The cooled amalgam is then pounded with the white of egg, and forms a liquid mass which may be applied with a brush.

Amalgams for Silvering Glass Globes, &c.

No. 1.—	Lead (pure)	1
	Tin	1
	Bismuth	1
	Mercury	1
		4 parts.

No. 2.—Lead	. . . .	1
Tin	. . . .	1
Bismuth	. . . .	1
Mercury	. . . .	2
		5 parts.

The lead and tin are to be melted first, after which bismuth is added. The drosses are removed, and mercury is poured into the compound, which is perfectly stirred. Leaves of Dutch gold are sometimes introduced into the mixture, according to the color which it is required to impart to the globes.

An *alloy for tinning* various utensils is made of from 6 to 8 parts of tin, and 1 part of iron.

We have already said that zinc has been employed for similar purposes. The galvanoplastic processes make it easy to deposit zinc, tin, lead, &c., upon iron or copper. We shall not linger on these applications, which do not belong to the subject of alloys.

Amalgam of Cadmium and Tin for Dentists.

Tin	. . . .	2
Cadmium	. . . .	1
		3

The two metals are melted together, and the button obtained is filed with a rasp. The metallic powder is then dissolved in a large quantity of mercury, the excess of which is expressed through a chamois leather. The friable mass thus obtained is kneaded in the fingers, and soon becomes soft and homogeneous. This paste, which rapidly hardens, is employed for filling teeth, and is also very serviceable as a hermetic luting for glass instruments, &c.

The following process, recommended by Mr. Boetger, is more rapid:—

As soon as the portions of cadmium and tin have been melted in an iron ladle, a certain portion of hot mercury is added to the mass, which is pounded and worked in an iron mortar until it has acquired a soft and butter-like consistency.

Alloy of Mr. Bibra for Small Casts.

Bismuth	6
Tin	3
Lead	13
	22

These metals are melted in a crucible or iron ladle, cast into ingots, and remelted before being employed. This alloy, which is nearly as fusible as that of Rose (bismuth 3, tin 1, lead 1), is harder, without being brittle or presenting a crystalline fracture. If the casts are wet with diluted nitric acid, then rinsed in water, and lastly rubbed with a woollen rag, the projecting parts become bright, while the cavities acquire the dark gray appearance of antique objects. Without acid the color of the metal is a light gray.

The medals cast upon plaster of Paris succeed so well that the finest and most delicate letters or lines, which, on the original piece, could be perceived with a magnifying glass only, become at once apparent to the naked eye. As the cost of bismuth is a great deal higher than that of tin, and especially that of lead, we may yet retain a good alloy by increasing the proportion of lead and diminishing that of bismuth.

This alloy may be useful in the manufacture of rollers and plates for calico printing.

The *alloy of Mr. Gersnein,* for making a soft mastic for uniting glass, chinaware, &c., becomes so hard after a certain lapse of time (8 to 10 hours), that it may be polished the same as silver or brass.

The copper employed is that obtained by precipitation. This copper is ground with concentrated oil of vitriol in a porcelain mortar, and then for from 25 to

35 parts of copper 65 to 70 parts in weight of mercury are gradually added. When the copper is entirely amalgamated, it is washed with boiling water, in order to remove the sulphuric acid, and then allowed to rest. This amalgam is unacted upon by the weak acids, alcohol, ether, or boiling water. Whenever it is desired to employ it as a mastic, it is always easy to bring it back to a soft and plastic state, by heating it up to about 375° C. and triturating it in a mortar until it has become as soft as wax.

If, in this state, it is put between two surfaces free from oxides, grease, &c., it unites them so thoroughly, that the pieces appear as if they had never been soldered. This copper amalgam has been employed by some dentists for filling teeth.

Alloy for roller scrapers:—

Copper	81.5
Zinc	10.5
Tin	8.
	100.0

This composition for the scrapers (sometimes called doctors, or ductors), intended to remove the surplus of colors from the calico-printing rollers, appears to possess the maximum of hardness and toughness for this purpose. On the other hand, acids rapidly destroy the scrapers made of an alloy of copper, tin, and zinc. For many years past, a combination which will possess, at the same time, elasticity and softness, hardness and flexibility, without being sensibly attacked by chemical reagents, has been a desideratum. The *Société Industrielle de Mulhouse* has offered a premium for such a discovery, which has not been yet awarded, because, as we believe, nothing has been invented which is to be preferred to the alloys made within the above limits.

Violet alloy, susceptible of a fine polish:—

Copper	75
Antimony	25
	100

This compound is brittle, without well-known uses, and more fusible than copper.

Amalgam for electrical machines:—

Zinc	1
Tin	1
Mercury	2
	4

This amalgam is employed, either in powder, or incorporated with grease.

Liquid for amalgamating the zinc of galvanic batteries:—

This liquid was experimented upon by Ruhmkorf. A few seconds of immersion are sufficient for amalgamating the most worn-out zinc. It is made by dissolving, with the aid of heat, 200 grammes of mercury in 100 grammes of aqua regia. When the solution is completed, 1000 grammes of hydrochloric acid are added to it.

Note by the Author.—Notwithstanding the innumerable researches which we have made in order to give a complete description of the useful alloys, it is probable, and even sure, that many interesting combinations have escaped our attention. Therefore, we shall welcome all communications and corrections on this subject, which our readers may have the kindness to address to us, in order thus to improve a future edition, if, as we hope, from the practical character and usefulness of a work of this kind, our book is to be printed again.

TABLES

SHOWING THE

RELATIVE VALUES OF FRENCH AND ENGLISH WEIGHTS AND MEASURES, &c.

Measures of Length.

Millimetre	=	0.03937	inch.
Centimetre	=	0.393708	"
Decimetre	=	3.937079	inches.
Metre	=	39.37079	"
"	=	3.2808992	feet.
"	=	1.093633	yard.
Decametre	=	32.808992	feet.
Hectometre	=	328.08992	"
Kilometre	=	3280.8992	"
"	=	1093.633	yards.
Myriametre	=	10936.33	"
"	=	6.2138	miles.
Inch ($\frac{1}{36}$ yard)	=	2.539954	centimetres.
Foot ($\frac{1}{3}$ yard)	=	3.0479449	decimetres.
Yard	=	0.91438348	metre.
Fathom (2 yards)	=	1.82876696	"
Pole or perch ($5\frac{1}{2}$ yards)	=	5.029109	metres.
Furlong (220 yards)	=	201.16437	"
Mile (1760 yards)	=	1609.3149	"
Nautical mile	=	1852	"

Superficial Measures.

Square millimetre	=	$\frac{1}{645}$	square inch.
" "	=	0.00155	" "
" centimetre	=	0.155006	" "
" decimetre	=	15.50059	" inches.
" "	=	0.107643	" foot.
" metre or centiare	=	1550.05989	" inches.
" " "	=	10.764299	" feet.
" " "	=	1.196033	" yard
Are	=	1076.4299	" feet.
"	=	119.6033	" yards.
"	=	0.098845	rood.
Hectare	=	11960.3326	square yards.
"	=	2.471143	acres.
Square inch	=	645.109201	square millimetres.
" "	=	6.451367	" centimetres
" foot	=	9.289968	" decimetres.
" yard	=	0.836097	" metre.
" rod or perch	=	25.291939	" metres.
Rood (1210 sq. yards)	=	10.116775	ares.
Acre (4840 sq. yards)	=	0.404671	hectare.

Measures of Capacity.

Cubic millimetre	=	0.000061027	cubic inch.
" centimetre or millilitre	=	0.061027	" "
10 " centimetres or centilitre	=	0.61027	" "
100 " " " decilitre	=	6.102705	" inches.
1000 " " " litre	=	61.0270515	" "
" " " " "	=	1.760773	imp'l pint.
" " " " "	=	0.2200967	" gal'n.
Decalitre	=	610.270515	cubic inches.
"	=	2.2009668	imp. gal'ns.
Hectolitre	=	3.531658	cubic feet.
"	=	22.009668	imp. gal'ns.
Cubic metre or stere or kilolitre	=	1.30802	cubic yard.
" " "	=	35.3165807	" feet.
Myrialitre	=	353.165807	" "

Cubic inch	=	16.386176	cubic centimetres.
" foot	=	28.315312	" decimetres.
" yard	=	0.764513422	" metre.

American Measures.

Winchester or U.S. gallon (231 cub.in.)	=	3.785209	litres.
" " bushel (2150.42 cub. in.)	=	35.23719	"
Chaldron (57.25 cubic feet)	=	1621.085	"

British Imperial Measures.

Gill	=	0.141983	litre.
Pint ($\frac{1}{8}$ gallon)	=	0.567932	"
Quart ($\frac{1}{4}$ gallon)	=	1.135864	"
Imperial gallon (277.2738 cub. in.)	=	4.54345797	litres.
Peck (2 gallons)	=	9.0869159	"
Bushel (8 gallons)	=	36.347664	"
Sack (3 bushels)	=	1.09043	hectolitre.
Quarter (8 bushels)	=	2.907813	hectolitres.
Chaldron (12 sacks)	=	13.08516	"

Weights.

Milligramme	=	0.015438395	troy grain.
Centigramme	=	0.15438395	" "
Decigramme	=	1.5438395	" "
Gramme	=	15.438395	" grains.
"	=	0.643	pennyweight.
"	=	0.0321633	oz. troy.
"	=	0.0352889	oz. avoirdupois.
Decagramme	=	154.38395	troy grains.
"	=	5.64	drachms avoirdupois.
Hectogramme	=	3.21633	oz. troy.
"	=	3.52889	oz. avoirdupois.
Kilogramme	=	2.6803	lbs. troy.
"	=	2.205486	lbs. avoirdupois.
Myriagramme	=	26.803	lbs. troy.
"	=	22.05486	lbs. avoirdupois.

Quintal metrique	=	100 kilog.	=	220.5486	lbs. avoirdupois.
Tonne	=	1000 kilog.	=	2205.486	" "

Different authors give the following values for the gramme:—

Gramme	=	15.44402	troy grains.
"	=	15.44242	"
"	=	15.4402	"
"	=	15.433159	"
"	=	15.43234874	"

AVOIRDUPOIS.

Long ton = 20 cwt. = 2240 lbs.	=	1015.649	kilogrammes.
Short ton (2000 lbs.)	=	906.8296	"
Hundred weight (112 lbs.)	=	50.78245	"
Quarter (28 lbs.)	=	12.6956144	"
Pound = 16 oz. = 7000 grs.	=	453.4148	grammes.
Ounce = 16 dr'ms. = 437.5 grs.	=	28.3375	"
Drachm = 27.344 grains	=	1.77108	gramme.

TROY (PRECIOUS METALS).

Pound = 12 oz. = 5760 grs.	=	373.096	grammes.
Ounce = 20 dwt. = 480 grs.	=	31.0913	"
Pennyweight = 24 grs.	=	1.55457	gramme.
Grain	=	0.064773	"

APOTHECARIES' (PHARMACY).

Ounce = 8 drachms = 480 grs.	=	31.0913	gramme.
Drachm = 3 scruples = 60 grs.	=	3.8869	"
Scruple = 20 grs.	=	1.29546	gramme.

CARAT WEIGHT FOR DIAMONDS.

1 carat = 4 carat grains = 64 carat parts.
" = 3.2 troy grains.
" = 3.273 " "
" = 0.207264 gramme
" = 0.212 "
" = 0.205 "
Great diversity in value.

Proposed Symbols for Abbreviations.

M—myria — 10000	Mm	Mg	Ml	
K—kilo — 1000	Km	Kg	Kl	
H—hecto — 100	Hm	Hg	Hl	Ha
D—deca — 10	Dm	Dg	Dl	Da
Unit — 1	metre—m	gramme—g	litre—l	are—a
d—deci — 0.1	dm	dg	dl	da
c—centi — 0.01	cm	cg	cl	ca
m—milli — 0.001	mm	mg	ml	

Km = Kilometre. Hl = Hectolitre. cg = centigramme. c. cm = $\overline{cm}^3$ = cubic centimetre. $\overline{dm}^2$ = sq. dm = square decimetre. Kgm = Kilogrammetre. Kg° = Kilogramme degree.

Celsius or Centigrade.	Fahrenheit.	Réaumur.
— 15°	+ 5°	— 12°
— 10	+ 14	— 8
— 5	+ 23	— 4
0 melting	+ 32	ice 0
+ 5	+ 41	+ 4
+ 10	+ 50	+ 8
+ 15	+ 59	+ 12
+ 20	+ 68	+ 16
+ 25	+ 77	+ 20
+ 30	+ 86	+ 24
+ 35	+ 95	+ 28
+ 40	+104	+ 32
+ 45	+113	+ 36
+ 50	+122	+ 40
+ 55	+131	+ 44
+ 60	+140	+ 48
+ 65	+149	+ 52
+ 70	+158	+ 56
+ 75	+167	+ 60
+ 80	+176	+ 64
+ 85	+185	+ 68
+ 90	+194	+ 72
+ 95	+203	+ 76
+100 boiling	+212	water + 80
+200	+392	+160
+300	+572	+240
+400	+752	+320
+500	+932	+400

1° C. $= 1^\circ.8$ Ft. $= \frac{9}{5}^\circ$ Ft. $= 0^\circ.8$ R. $= \frac{4}{5}^\circ$ R.

1° C. $\times \frac{9}{5} = 1^\circ$ Ft. 1° Ft. $\times \frac{5}{9} = 1^\circ$ C. 1° R. $\times \frac{9}{4} = 1^\circ$ Ft.

1° C. $\times \frac{4}{5} = 1^\circ$ R. 1° Ft. $\times \frac{4}{9} = 1^\circ$ R. 1° R. $\times \frac{5}{4} = 1^\circ$ C.

Calorie (French) = unit of heat } English.
= kilogramme degree }

It is the quantity of heat necessary to raise 1° C. the temperature of 1 kilogramme of distilled water.

Kilogrammetre = Kgm = the power necessary to raise 1 kilogramme, 1 metre high, in one second. It is equal to $\frac{1}{75}$ of a French horse power. An English horse power = 550 foot pounds, while a French horse power = 542.7 foot pounds.

Ready-made Calculations.

No. of units.	Inches to centimetres.	Feet to metres.	Yards to metres.	Miles to Kilometres.	Millimetres to inches.
1	2.53995	0.3047945	0.91438348	1.6093	0.03937079
2	5.0799	0.6095890	1.82876696	3.2186	0.07874158
3	7.6199	0.9143835	2.74315044	4.8279	0.11811237
4	10.1598	1.2197680	3.65753392	6.4373	0.15748316
5	12.6998	1.5239724	4.57191740	8.0466	0.19685395
6	15.2397	1.8287669	5.48630088	9.6559	0.23622474
7	17.7797	2.1335614	6.40068436	11.2652	0.27559553
8	20.3196	2.4383559	7.31506784	12.8745	0.31496632
9	22.8596	2.7431504	8.22945132	14.4838	0.35433711
10	25.3995	3.0479450	9.14383480	16.0930	0.39370790

No. of units.	Centimetres to inches.	Metres to feet.	Metres to yards.	Kilometres to miles.	Square inches to square centimetres.
1	0.3937079	3.2808992	1.093633	0.6213824	6.45136
2	0.7874158	6.5617984	2.187266	1.2427648	12.90272
3	1.1811237	9.8426976	3.280899	1.8641472	19.35408
4	1.5748316	13.1235968	4.374532	2.4855296	25.80544
5	1.9685395	16.4044960	5.468165	3.1069120	32.25680
6	2.3622474	19.6853952	6.561798	3.7282944	38.70816
7	2.7559553	22.9662944	7.655431	4.3496768	45.15952
8	3.1496632	26.2471936	8.749064	4.9710592	51.61088
9	3.5433711	29.5280928	9.842697	5.5924416	58.06224
10	3.9370790	32.8089920	10.936330	6.2138240	64.51360

No. of units.	Square feet to sq. metres.	Sq. yards to sq. metres.	Acres to hectares.	Square centimetres to sq. inches.	Sq. metres to sq. feet.
1	0.0929	0.836097	0.404671	0.155	10.7643
2	0.1858	1.672194	0.809342	0.310	21.5286
3	0.2787	2.508291	1.204013	0.465	32.2929
4	0.3716	3.344388	1.618684	0.620	43.0572
5	0.4645	4.180485	2.023355	0.775	53.8215
6	0.5574	5.016582	2.428026	0.930	64.5858
7	0.6503	5.852679	2.832697	1.085	75.3501
8	0.7432	6.688776	3.237368	1.240	86.1144
9	0.8361	7.524873	3.642039	1.395	96.8787
10	0.9290	8.360970	4.046710	1.550	107.6430

No. of units.	Square metres to sq. yards.	Hectares to acres.	Cubic inches to cubic centimetres.	Cubic feet to cubic metres.	Cubic yards to cubic metres.
1	1.196033	2.471143	16.3855	0.02831	0.76451
2	2.392066	4.942286	32.7710	0.05662	1.52902
3	3.588099	7.413429	49.1565	0.08494	2.29354
4	4.784132	9.884572	65.5420	0.11325	3.05805
5	5.980165	12.355715	81.9275	0.14157	3.82257
6	7.176198	14.826858	98.3130	0.16988	4.58708
7	8.372231	17.298001	114.6985	0.19819	5.35159
8	9.568264	19.769144	131.0840	0.22651	6.11611
9	10.764297	22.240287	147.4695	0.25482	6.88062
10	11.960330	24.711430	163.8550	0.28315	7.64513

No. of units.	Cubic centimetres to cubic inches.	Litres to cubic inches.	Hectolitres to cubic feet.	Cubic metres to cubic feet.	Cubic metres to cubic yards.
1	0.06102	61.02705	3.5317	35.31659	1.30802
2	0.12205	122.05410	7.0634	70.63318	2.61604
3	0.18308	183.08115	10.5951	105.94977	3.92406
4	0.24411	244.10820	14.1268	141.26636	5.23208
5	0.30514	305.13525	17.6585	176.58295	6.54010
6	0.36617	366.16230	21.1902	211.89954	7.84812
7	0.42720	427.18935	24.7219	247.21613	9.15614
8	0.48823	488.21640	28.2536	282.53272	10.46416
9	0.54926	549.24345	31.7853	317.84931	11.77218
10	0.61027	610.27050	35.3166	353.16590	13.08020

No. of units.	Grains to grammes.	Ounces avoir. to grammes.	Ounces troy to grammes.	Pounds avoir. to kilogrammes.	Pounds troy to kilogrammes.
1	0.064773	28.3375	31.0913	0.4534148	0.373096
2	0.129546	56.6750	62.1826	0.9068296	0.746192
3	0.194319	85.0125	93.2739	1.3602444	1.119288
4	0.259092	113.3500	124.3652	1.8136592	1.492384
5	0.323865	141.6871	155.4565	2.2670740	1.865480
6	0.388638	170.0250	186.5478	2.7204888	2.238576
7	0.453411	198.3625	217.6391	3.1739036	2.611672
8	0.518184	226.7000	248.7304	3.6273184	2.984768
9	0.582957	255.0375	279.8217	4.0807332	3.357864
10	0.647730	283.3750	310.9130	4.5341480	3.730960

No. of units.	Long tons to tonnes of 1000 kilog.	Pounds per square inch to kilogrammes per square centimetre.	Grammes to grains.	Grammes to ounces avoir.	Grammes to ounces troy.
1	1.015649	0.0702774	15.438395	0.0352889	0.0321633
2	2.031298	0.1405548	30.876790	0.0705778	0.0643266
3	3.046947	0.2108322	46.315185	0.1058667	0.0964899
4	4.062596	0.2811096	61.753580	0.1411556	0.1286532
5	5.078245	0.3513870	77.191975	0.1764445	0.1608165
6	6.093894	0.4216644	92.630370	0.2117334	0.1929798
7	7.109543	0.4919418	108.068765	0.2470223	0.2251431
8	8.125192	0.5622192	123.507160	0.2823112	0.2573064
9	9.140841	0.6324966	138.945555	0.3176001	0.2894697
10	10.156490	0.7027740	154.383950	0.3528890	0.3216330

No. of units.	Kilogrammes to pounds avoirdupois.	Kilogrammes to pounds troy.	Metric tonnes of 1000 kilog to long tons of 2240 pounds.	Kilog. per square millimetre to pounds per square inch.	Kilog. per square centimetre to pounds per square inch.
1	2.205486	2.6803	0.9845919	1422.52	14.22526
2	4.410972	5.3606	1.9691838	2845.05	28.45052
3	6.616458	8.0409	2.9537757	4267.57	42.67578
4	8.821944	10.7212	3.9383676	5690.10	56.90104
5	11.027430	13.4015	4.9229595	7112.63	71.12630
6	13.232916	16.0818	5.9075514	8535.15	85.35156
7	15.438402	18.7621	6.8921433	9957.68	99.57682
8	17.643888	21.4424	7.8767352	11380.20	113.80208
9	19.849374	24.1227	8.8613271	12802.73	128.02734
10	22.054860	26.8030	9.8459190	14225.26	142.25260

INDEX.

www.ingramcontent.com/pod-product-compliance
Lightning Source LLC
LaVergne TN
LVHW020241110826
845151LV00003B/999

* 9 7 8 1 4 2 5 5 2 9 9 2 5 *